Focus on AUSTRALIAN BIRDS

Focus on

AUSTRALIAN BIRDS

Peter Antill-Rose & Associates Pty Ltd

8/10 Anella Avenue
Castle Hill 2154
Australia

First published in Australia in 1991 by:

Peter Antill-Rose and Associates Pty Ltd
8/10 Anella Avenue
Castle Hill 2154
New South Wales
Australia

Publisher: Allan Cornwell
Editor: Judi Stansfield

Designed by Small Back Room Productions, Upwey, Victoria
Typeset by Bandaid Productions, Fitzroy, Victoria
Colour Separations by Scanagraphix Pty Ltd, Brunswick, Victoria
Printed in Australia by Griffin Press, Netley, South Australia.

ISBN 1 86282 088 0

BIRDS IN AUSTRALIA

Australia is well represented by bird life. We have over 700 different species of birds, of which about 600 breed and live permanently on land. Of the remaining, about 80 different kinds breed in swamps and waterside environments, mainly in the coastal and sub-coastal regions, about 30 kinds of sea birds breed on the coasts and on off shore islands, and in addition, there are many regular non-breeding migrants. Of the 600 known native resident species, nearly 400 are endemic, that is found nowhere else. Our dominant species — the cockatoos and parrots, kingfishers, bowerbirds, honeyeaters, nocturnal frogmouths and robin-whistlers are almost completely confined to the Australian continent.

The origins of Australia's bird life are still very much a mystery. At one time it was believed that Australia had separated from the Afro-Asian land masses during the Mesozoic age, that is prior to the existence of modern bird species. It was held that bird life gradually colonised our country by spreading from South-East Asia by way of Indonesia and New Guinea into Northern Australia. These birds were thought to be the ancestors of the main group of endemic bird species such as the mound-builders, the parrots, the lyrebirds, the honeyeaters, the mudnesters, the butcherbirds, and the bowerbirds and birds-of-paradise. This theory was highly regarded and supported by the fact that so many "Old World" bird families, such as flamingoes, the true wrens, woodpeckers and vultures, are entirely absent from Australia. It was assumed that they had been incapable of making the migratory journey across the sea.

When fossil remains of four species of flamingoes were found in the Lake Eyre Basin, it was realised that they had once existed in Australia but had become extinct. If this were true of flamingoes it was possible other old world birds had also lived here, thus casting doubt on the "colonisation" theory. Recent DNA studies also are beginning to make this theory even more doubtful as they show that the genetic links between Australian bird groups and their apparent Asian counterparts are tenuous, to say the least.

An alternative theory which may help to explain the origins of Australia's modern bird life, but which is in no way certain, is that some 100 million years ago Australia was part of a huge, southern supercontinent called Gondwanaland which also included Africa, India, Madagascar, Antarctica, New Zealand and South America. Some of our land birds are thought to have come from Gondwanaland. These include the Cassowary, the Emu, the mound-builders, the Plains Wanderer, the parrots, the Pied Goose and maybe even the pigeons, cuckoos and rails. When the supercontinent broke apart Australia was separated from the other land masses of the world but by then it had the foundation of its own bird life. Subsequent migration has increased the number of species found in Australia.

Habitats

Today, bird life is found almost everywhere in Australia from arid scrubland to the rich rainforest areas. Each species of bird has its own preferred habitat or environmental area and the distances it is willing to travel to forage for food also varies. Some birds range through a variety of habitats without ever moving far. Others wander more widely, even migrating, either in an attempt to follow the food they prefer, or to cope with erratic changes in their food supplies. Some birds live in similar habitats in different places at different times of the year, while others are sedentary, living all year round in one habitat.

The range of habitats in which Australian birds can be found can be conveniently broken up into different areas: Rainforests, Sclerophyll Forests, Woodlands, Mallee and Mulga, Shrub Steppes, Grasslands, Heathlands, Mangroves, Wetlands, Shores and Mudflats, and Islands. Each area is marked by a particular climate,

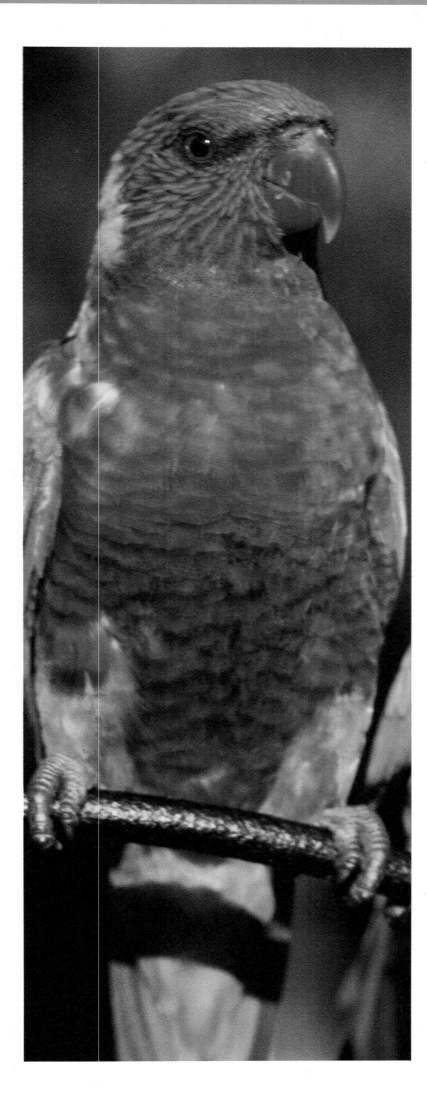

type of vegetation and animal and bird life. The relationships of all these things with each other is vital.

The rainforest is where the highest proportion of resident bird species are to be found. The forests are dense, dark and evergreen, with a variety of trees growing tall and close together to form a canopy that keeps out sunlight but lets the high year-round rainfall through. About 140 species of birds live in Australian rainforests, over 60 of them, like the cassowary and bowerbird, being virtually confined to this habitat.

There are three main forms of rainforest in Australia. One is the tropical rainforest which ranges along eastern Cape York. Most of the birds in this area are New Guinean in origin and are found nowhere else in Australia. The most widespread rainforest is subtropical which occurs as far north as the tablelands of Cairns-Cooktown, and extends south to the Illawarra area in New South Wales. Here in this sub-tropical rainforest, the Australian Brush Turkey is found. The third type of rain forest, the temperate, is limited to Tasmania and wet areas of southeastern Australia where only a few species such as the Pink Robins, and Rufous Scrub-birds are to be found.

Sclerophyll forest areas are dominated by tall to medium-high eucalypts which form a fairly open canopy. In the coastal and tableland districts of the southeast and southwest the forests are wet, with a dense understorey of tall shrubs, small trees and tree ferns. Here is the home of the Gang Gang Cockatoo and the Powerful Owl. On the less fertile soils of the coast and tablelands are the dry sclerophyll forests where the canopy is more open and the shrubs and grasses are more abundant. Here Scarlet Robins live.

Woodland areas resemble parklands in appearance with small to medium-high trees with large branch and leaf crowns and grassy covered soil beneath. There are three main types of woodland in Australia: the tropical eucalypt of the North, the temperate eucalypt of the southeast and southwest, and the arid to semi-arid woodlands around central Australia. Together the tropical and temperate woodland habitats support over half of Australia's land birds. Among these are the Sulphur Crested Cockatoo, White-cheeked Rosella and the White-plumed Honeyeater.

Mulga occurs patchily across the alluvial flatlands and tablelands of inland Australia where rainfall is usually less than 30 centimetres a year. Southward it reaches the fringes of the Mallee, and northward it ends at the Great Sandy Desert. Mallee and Mulga are marked by a dense growth of low trees, such as Melaleuca, usually only four to six metres high and with an almost continuous canopy, with hummocks of spinifex grass growing in deep sand areas. About 160 species of bird are found naturally in the mallee, including the yellow-rumped race of Spotted Pardalotes, and the Mallee Ringneck, while 80-90 species of birds are regularly seen in mulga regions.

On the red and grey loam and gibber plains of the southern inland, including the Nullarbor, where annual rainfall is less than 50 centimetres, grow the low bluebush and saltbush which form the treeless shrub steppe. Here, in this steppe region, some 50 species of birds live, and a few of which are specialised to live in it alone. These include the Cinnamon and Nullarbor Quail-thrushes and the Gibber-bird.

The grasslands of western Queensland, the Northern Territory and Western Australia are home to about 100 species of birds including the Spinifex Pigeon and the Spinifexbird. In these regions where there are usually no trees, hummock and spinifex grasses grow both in the sandridged deserts and on the stony, well-drained ranges. The spiny hummocks of grass grow barely 3-6 centimetres high.

Although there are a variety of heathlands, all are dense, treeless areas of prickly-leaved shrubs, usually less than two metres high. While coastal heaths are most widespread, heathlands also occur in

alpine areas, on sand plains and near mallee areas. Coastal heathlands are alive with birds though only a very few differing species actually live there. The dense cover and flowering shrubs provide nectar and pollen and attract insects. Here the Ground Parrots, honeyeaters and thornbills are most characteristic. In total about 80 species of birds regularly use heathlands.

Mangroves grow in muddy bays and estuaries of Northern Australia, in patches in the west, and in places along Westernport Bay in Victoria. Because the mud chokes oxygen from the water, the roots of the mangroves grow above the mud to obtain supplies of the vital gas from the air. Here we find an important habitat for up to 60-70 different species of mud and foliage feeding birds. About 20 species are characteristic of them and 12-13 of these live nowhere else in Australia. Here in the mangroves you may catch a glimpse of the Striated Heron, or the Yellow White-eye.

Along the shoreline as well as along the estuaries of rivers and the edges of lakes, water birds can be found wading in the mud and sand or on reefs probing the ground for food. Plovers, sandpipers and curlews congregate on their favourite feeding grounds or band together to breed in areas such as the Eighty Mile Beach in northwestern Australia, in the Gulf of Carpentaria, and in the Coorong in South Australia.

Islands off the coast of Australia are important breeding grounds for about 26 species of sea birds. On islands like the coral cays of the Barrier Reef, the Albrolhos group of Western Australia, Cabbage Tree Island of Port Stephens, and Phillip Island in Westernport Bay, Black-winged Petrels, terns, Gould's Petrels and Little Penguins, respectively, breed.

BIRDS IN GENERAL

Bird Behaviour

Flying has always been one of man's obsessions. To conquer the skies and experience the ultimate freedom that flying promises has been one of our dreams. For this reason birds have always been of fascination to us. They capture our attention with their ability to fly, to soar on currents of air, to twist and turn, to dive on prey from great heights, and to hover as if suspended in air on a thread. For us, the bird is truly the ultimate in flying machines.

Yet it is not a bird's ability to fly which separates it from other living things. There are in fact bats and insects that can fly and some species of birds which cannot. What truly distinguishes birds from other animals is their feathers.

Although fossils give us proof that birds evolved from reptiles it is the presence of feathers which most clearly differentiated them from the reptile family.

The earliest known fossils of a bird are those of Archaeopteryx that lived in Germany about 150 million years ago. Many features of Archaeopteryx were similar to those of reptiles: teeth, a long tail with 20 vertebrae, three fingers with a claw, horny scales on the legs and toes, and particularly reptilian skeletal features. It also had some very bird like features: streamlined pubic bones suitable for flight, joined collarbones in a wish bone shape, and an indication that it laid hard-shelled eggs. However, the one feature that clearly showed it was not a reptile was feathers. The wings of early reptiles that achieved flight were made up of a membrane, such as is present in bats, and not flight feathers.

Although it is feathers that primarily distinguish birds from other animals, all birds conform to a basic structure and have a number of similar characteristics. Their forelimbs have been modified into wings forcing the backlegs to support the body and give the bird the upright, two legged stance it has. Their skeletons have developed for flight making their bodies streamlined. Their bones are strong, though at the same time porous and hollow to keep the bird light. They all have a beak, specialised for feeding.

They are warm-blooded and lay hard-shelled eggs which one or both parents incubate. And, of course, their bodies are covered in feathers for flight, for heat conservation, for waterproofing, for camouflage, and for display. These characteristics allow us to distinguish over 8,600 different species in the world that are birds and not animals. Differentiating them from each other, however, is another matter.

We tend to rely on the visible external characteristics — a bird's size, shape and colour — in telling one bird from another. A closer examination shows us that it is their behaviour — their locomotion, feeding, grooming, communication, territorial defence, courtship and breeding habits — that gives us a much clearer indication of a species or genus of bird.

On land birds move in different ways. Some "walk" moving one foot after another, while others stretch out and run. Some flightless birds, such as Emus, have lost their ability to escape by flying, and can run at speeds up to 45km. Other birds hop around the ground with both feet together, which is the most efficient way though also very energy consuming.

The way a bird chooses to move across land, and the ease with which it does this, is largely dependent on the structure of its leg and foot, and these are in turn related to its habitat and sometimes its food. For instance, birds which wade in shallow waters in search of food require long, stilt-like legs, whereas birds who swim in search of food have webbed feet or paddle-shaped toes.

Birds use flight for different purposes varying from hour to hour, and day to day. Some are travelling from one habitat to another using the skies to travel vast distances. Some are searching for food and water, or using flight as a means of capturing their prey. Some engage in display flights, as the breeding season approaches, as a way of attracting a mate. At times a bird takes flight merely to escape danger.

Because a bird spends such a large amount of time searching for food, they have specialised methods of finding it. They all have bills and the variations in shape and form they take illustrates the extensive range of foods birds can eat. Each bill is developed to fit a particular habitat and method of food gathering. Falcons tear the flesh of their prey apart with their strong, hooked beak, while a cormorant will grasp a fish in its bill. The Spoonbill uses its spatula-like bill to sift its food from other matter in shallow water, while herons will spear fish with their dagger-shaped bills. Wading birds will walk through the shallows and use their bills to probe sand and silt for food. In forest areas some birds will rake over the litter of leaves on the forest floor with their beaks to expose food, while others cling to trees using their bills to probe the soft bark for insects. Parrots have yet another variation: they use their strong bills as a third leg to help them climb.

The need to establish territory is another very important characteristic of bird behaviour. Birds appear to have some instinctive code which assures that aggression is only shown while another bird is within its own defended territory. The moment a bird realises it is trespassing in someone else's it will retreat. Most birds signal their territorial right by issuing frightening calls, and through threatening aggressive behaviour, such as spreading their wings, raising their crests, or puffing out their feathers. In most cases there is little need for them to carry out the "threat" implied by their actions.

Territory is generally established by the male for the purpose of courtship and breeding. Attracting a mate is a specialist art. Some males have brightly coloured plumage and special features such as crests, plumes or wattles, while song birds rely heavily on courtship tunes which they sing to a potential mate. Many have specialised displays either on the ground or in the air. On the ground these can involve any number of actions including dancing, bobbing, bowing, bill fencing, wing flapping, preening, and feeding each other. Each species has its own courtship rituals which lead to mating, and

each species has its own ways of coping with reproduction and raising its young.

Most birds prepare some sort of nest in which to lay and incubate their eggs. For some birds, like terns, this can be as simple as scraping a depression in the ground, whereas for other birds, such as the Mistletoe-bird it can involve the construction of a special shaped nest woven of down and spider web. The basic functions of the nest are to cradle the eggs and the young hatchlings and keep them relatively safe from harm. Nests are generally built from vegetable matter such as sticks, seaweed, grass, bark and leaves. Sometimes other materials like feathers, wool, hair and cobwebs are used, and some birds will even "paint" the inside of nests with the juice of berries, or their own saliva. Nests are also made in a variety of shapes and locations: the Mistletoe-bird builds a pear-shaped nest; the Robin has its tiny cup; the Grebe's nest is a platform of reeds and rushes which floats on the water; the Rainbow Bee-eater's nest chamber is at the end of a tunnel in a sandy bank; and the Mallee-fowl buries its eggs in a mound of warm, fermenting, vegetable matter.

The egg is hatched when the young chick breaks through the shell with a special egg tooth on its bill. This tooth disappears shortly after hatching. There are two types of young: the precocial which, within a few hours of hatching, can run around by itself, and the altricial, which is totally dependent and nest bound for some days. The Malleefowl chick is an excellent example of a precocial hatchling as he must fend for himself as soon as he has dug his way out of the mound. Most bird parents, however, attend their chicks after hatching, usually cleaning the nest, protecting the chicks, keeping them warm and feeding them. In some species of bird these tasks are the duties of one or both parents, while in other species large family groups assist.

CONSERVATION

Australia's birds are precious and we need to protect them. The best way to do this is to maintain their habitats. However, our record of doing this is not good. We have altered every avian habitat since European settlement began, and while we have not yet destroyed any completely, some, such as the subtropical rainforest and mallee, have been severely affected.

In the past we have grazed our stock on natural grassland, and let rabbits run loose on them to breed prolifically; we have mined the heathlands and polluted mangroves and mudflats with industrial wastes; we have drained swamps in an effort to regain land; we log forests daily to meet our wood consumption needs, or thin them out to graze extra cattle. When we alter or destroy a bird's habitat we limit its ability to survive. Many birds are restricted in their choice of habitat. The Malleefowl, for instance lives only in mallee areas and continual clearing and grazing of its habitat threatens its survival. Similarly exotic rainforest dwellers cannot survive the continued logging of their habitat and there is no alternative place for them to live.

To protect and maintain our bird population should be a major task for all of us. If a majority of Australians cared enough and sincerely believed our birds were worth conserving then perhaps we could work together to secure for them their rightful place: the forests, the open skies, and the seas.

Australian Pelican
Pelecanus conspicillatus

Other names: Spectacled Pelican.

Distribution: Throughout Australia where suitable habitats exist.

Notes: Pelicans have lived in Australia for a long time; their fossil remains suggesting some 30 to 40 million years. Living mostly in flocks, they frequent the estuaries of rivers, coastal mud-flats or rivers, and lakes of island areas.

The pelican is at home in the water. Flocks of them will swim in formation, herding fish into the shallow waters where the birds will then scoop them up in their distinctive, large bills. The water they trap simultaneously, is squeezed out through "valves" at the corners of their mouths. Both parents will feed their chicks with the food they collect in their bills, and the chicks respond by going into convulsions and collapsing after eating. Each convulsion lasts for about a minute and is thought to be an extreme form of "begging".

The pelican is a graceful bird both on the water and in the air, but by contrast is clumsy on land and in its attempts to take off.

Southern Cassowary
Casuarius casuarius

Other names: Double-wattled Cassowary, Australian Cassowary

Distribution: Restricted to tropical rainforests in north eastern Queensland where it is fairly common.

Notes: The Southern Cassowary is a solitary bird, living much of its life alone. It scares off intruders by "rumbling" at them, very much in the manner of an approaching truck. Cassowaries will only pair when mating. After the female has laid her eggs she will leave them for the male to incubate and raise. The male will sit the eggs for about two months and look after the chicks for about nine months. Cassowaries can be very aggressive, especially when defending their young, and are armed with a large, straight spike on the inside toe of each foot which they will use in a fight.

Cassowaries inhabit tropical rainforest areas and feed on the fallen fruit from rainforest trees and vines. They have been known to raid gardens and orchards for food, will eat almost anything from fungi to dead rats, but strangely do not particularly like citrus fruit.

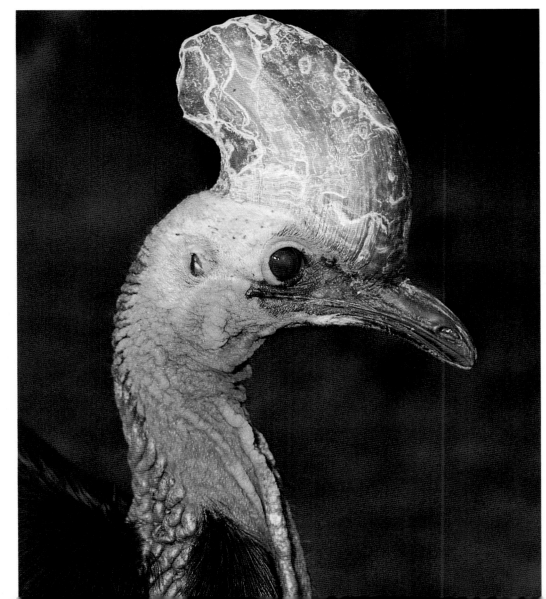

Black Swan
Cygnus atratus

Other Names: None

Distribution: Suitable parts of continental Australia latterly extending in north-east Tasmania; introduced into New Zealand.

Notes: Black Swans can be found in freshwater and briny swamps, rivers, estuaries and lakes in many parts of Australia. They live and breed freely in parks and on ornamental lakes even in major cities. They have a distinctive trumpeting call with which they hail each other, and also use the white on their outer flight feathers to signal flight. Then the flock patters across the water into the air on heavily beating wings. Black Swans do much of their flying at night.

Each year between September and February Black Swans moult, making them flightless. As a result many hundreds of them may congregate in the one breeding ground where they feed on aquatic plants and animals. Often where large colonies of the swan do breed together, nests are destroyed as the birds pilfer each others' nest materials. As eggs are deserted and scattered, neighbouring swans scrape the stray ones into their own nests and incubate them.

Australian Bustard
Ardeotis australis

Other names: Kori Bustard, Plains Turkey, Wild Turkey

Distribution: Formerly throughout inland Australia extending to coastal areas north of the Tropic of Capricorn; now much reduced and only numerous in central and northern Queensland and Kimberleys.

Notes: Once this large and striking bird could be encountered almost anywhere in open country, but now the Kori Bustard's numbers have been reduced by the expanding human population and by the effects of introduced animals such as sheep, rabbits and foxes.

One of the most spectacular features of this bird is the male's mating display. He struts about on a display ground with tail thrown forward over the back, long breast-feathers fanned, and pendulous white breast-sac lowered to the ground, uttering a low hollow roar and booming loudly.

Kori Bustards like open wooded grass plains. Their diet consists of grasses, the seeds and fruits of native plants, some insects, and small mammals and birds. They are very much a nomadic bird, their movements apparently governed by rainfall.

Emu
Dromaius novaehollandiae

Other Names: None

Distribution: Continental Australia generally, except the more closely settled coastal areas; extinct in Tasmania, King Island and Kangaroo Island.

Notes: Next to the Ostrich, the Emu is the largest of all birds, standing up to 2 metres in height. This large, flightless bird is a nomadic wanderer in most parts of Australia, often moving several hundred kilometres in a year. They have distinctively "shaggy" plumage because each feather base has two long plumes growing from it.

Emus are omnivorous, eating a variety of leaves, grasses, fruits and flowers of native plants, as well as insects. They tend to suffer severely during periods of drought as food sources become scarce.

The female Emu can lay up to 20 eggs in a clutch, though seven to eleven is the average. Immediately after laying her eggs, the female will wander off leaving the male to incubate them. He will sit them for eight weeks often losing 4 to 8 kilos in weight during this time. After hatching, the male will continue to rear the chicks for up to 18 months.

Black-Necked Stork
Xenorhynchus asiaticus

Other Names: Jabiru, Policeman-bird

Distribution: Along the north and east coasts, sometimes as far south as Sydney but less common in southern part of range. Rare vagrant to Victoria.

Notes: The Black-necked Stork is often called the Jabiru and this is sometimes thought to be the Aboriginal name for the bird. It is, however, a Portuguese word and other storks in South America and Africa are also referred to as Jabirus.

Unlike others of their species, the Black-necked Stork is not a social bird preferring to remain independent, or in a loose pair. Because adult birds tend to stay in pairs during the non-breeding season, it is thought they pair permanently, and use the same nests year after year. These they build on the tops of trees or large bushes and are substantial structures.

The Black-necks are freshwater foragers basing their diets on fish, reptiles, frogs and crabs. Occasionally they will run a few steps to catch food, but mostly they stand and wait, or stalk their prey, using their strong bills to grasp food. In flight the stork will often soar several hundred metres high, trailing its legs behind, but this is only after a series of running jumps and a slow flapping of the wings.

Cape Barren Goose
Cereopsis novaehollandiae

Other Names: Pig Goose

Distribution: Islands off southern coast of Australia, from Furneaux Group, Tasmania, to Recherche Archipelago Western Australia.

Notes: These distinctive geese are unique to Australia and have been hunted and eaten since they were first seen by George Bass in 1797. Now they are fully protected.

Cape Barren Geese live in pairs or small flocks, frequenting grasslands and swampy areas, where they feed on grasses and herbage of various kinds. The name Pig Goose refers to the grunt like noises they make; the male also has a high pitched trumpet. The male is also known for his intolerance of others and is fond of fighting to establish his territory as the breeding season approaches.

Cape Barren Geese probably mate for life and share the rearing of the young until they are about six weeks old. After breeding the adults moult and for a short time are completely flightless. They fly strongly with their necks outstretched but being such large birds have a long lumbering run to take off.

Magpie Goose
Anseranas semipalmata

Other Names: Pied Goose, Black and White Goose, Semipalmated Goose, Wild Goose

Distribution: Northern and eastern Australia on the east coast to about Rockhampton, Queensland.

Notes: The Magpie Goose was once found all down the east coast to Victoria and eastern South Australia, however, human settlement combined with a couple of bad droughts has made the bird extinct in these areas. Now it is only to be found in the swamps, lagoons, mangrove flats, lakes and rivers of the north, however, agricultural development and grazing of the swamp vegetation by the introduced water buffalo are now posing a threat to the geese there.

The Magpie Goose is a large noisy bird which has a highpitched honk. In flight and on the ground a male call is immediately answered by females. The bird is also distinguished by its bold markings, its half-webbed feet and the knob on its head. The knob is larger on a male than female, but its size also depends on the age of the bird. They feed on aquatic plants and animals in the shallows of waterways.

In the breeding season the male may mate with two females. Breeding pairs and trios build floating platforms of rushes and herbage on trampled down tussocks of spike-reed near the middle of the swamp. On this platform the courtship display begins. Shortly before the female lays, the birds make the platform more elaborate and substantial, and after the egg is laid they add more material to it to make a thick, deep nest-cup.

Rufous Fantail
Rhipidura rufifrons

Other Names: Wood Fantail, Rufous Flycatcher, Rufous-fronted Fantail

Distribution: Coastal northern and eastern Australia and adjacent ranges from the Kimberleys, Western Australia, to south western Victoria in wet forests and mangroves.

Notes: The Rufous Fantail appears slight in form and slow and fluttering in flight, however, this is deceptive because the bird travels great distances each year. Migrating birds travel alone, flitting low from shrub to shrub, and picking up beetles, bugs, weevils, mites and spiders as they go. Much of their travelling is done in twilight hours.

The Rufous Fantail frequents mainly dense, low cover in rainforests, especially tree-fern gullies. Wherever they are they forage within a metre or two of the ground. They launch out continuously in short sallying flights from one perch to another, raising their constantly fanned tails vertically for takeoff and tumbling, diving and twisting up and leaping among close foliage.

There are two distinct races of the Rufous Fantail in Australia: one in the mangroves around the north coast; the other in the rainforests and wet sclerophyll forests along the east coast.

Eastern Yellow Robin

Eopsaltria australis

Other Names: Bark Robin, Creek Robin, Yellow Bob

Distribution: Eastern Australia from Cooktown south to south eastern Australia. Distribution patchy.

Notes: The Eastern Robin is most commonly found in coastal mountain forests, perched sideways on a low vertical branch or sapling. The Eastern is the yellowest of all robins with the colour of their underneath feathers varying from citrine to bright yellow. The robins flit quietly from perch to perch, where they will sit and wait, their tails occasionally rising and wings flicking. Then they will fly on or dart quickly to the ground to pick up prey. Their diet includes ants, spiders, moths, grasshoppers, wasps and flies.

The female robin builds a cup nest of bark strips, fine twigs, moss, skeleton leaves and grass bound with cobweb. She may adorn it with hanging strips of camouflaging, grey bark. In this she will lay her eggs at 27 hour intervals while the male feeds her. The raising of the new chicks is a family affair with one or two additional helpers, usually previous young, assisting.

Satin Bowerbird

Ptilonorhynchus violaceus

Other Names: Satinbird

Distribution: Eastern Australia (coast and associated highlands) from the Atherton Tablelands south to the Seaview Ranges near Townsville, Queensland, and from about Rockhampton to the Otway Ranges, Victoria.

Notes: The male and female Satin Bowerbirds are perfect examples of how different the sexes of birds can be. While the female is a dull grey-green and brown colour, her male counterpart is entirely black with a soft, satiny purple-blue sheen. He is considered a rather beautiful bird.

The bowerbird gets its name as a result of the bower it builds for nuptial display. Bowers are neat avenues of twigs woven upright into walls aligned north and south. They are built by the male bird. The Satin Bowerbird has the habit of decorating the ends of his bower with flowers, feathers, berries, and man made objects, preferably blue in colour. When a female enters the bower, the male will display and then copulate. The female then leaves the bower and will raise her young completely independent of the male.

For most of the year Satin Bowerbirds live in rainforest and wet sclerophyll forests, but in Autumn and Winter they roam widely in flocks of up to 50 in search of fruit, shoots and insects, before returning to their breeding grounds to build bowers.

Regent Bowerbird

Sericulus chrysocephalus

Other Names: Regentbird, Australian Regentbird

Distribution: Eastern Australia, from Clarke Range near Mackay south to the Hawkesbury River, New South Wales

Notes: Like all Bowerbirds, the Regent builds a bower as part of its mating display, however, it seems to bring truth to the notion that the brighter the plumage, the less elaborate the bower. In fact, the Regents' bowers are such an uncommon sight, it is not certain that all males build a bower, and that several birds may actually share one.

The bowers they do build are constructed on a loose platform of sticks up to a metre in diameter. Its parallel walls are about 20cm long, up to 30cm high and about 9cm apart. The bird "paints" the inside walls of the bower yellow with saliva and juice from crushed berries, and decorates the floor with snail shells, berries and leaves.

Regent Bowerbirds live in subtropical rainforests in the mid and upper levels. They descend only to the forest floor for bower-making, display and mating. They feed on fruit and berries from native trees.

Tawny Frogmouth
Podargus strigoides

Other Names: Podargus, Mopoke

Distribution: Open forests and woodlands of eucalypts and acacias throughout Australia including Tasmania.

Notes: The Tawny Frogmouth is a master of disguise. At the hint of a disturbance it freezes simulating the bark of the branch, or a broken branch, so that it is difficult to identify the bird itself.

  Like other frogmouths, the Tawny is nocturnal. They are most active in the hours just after dark and preceding dawn. During this time they watch carefully from suitable perches for prey, mostly large insects, spiders and myriapods. The frogmouth is, however, at risk of being killed by cars at night as it hunts. A male and female Tawny Frogmouth will pair permanently, and will roost near each other, not only during the breeding season, but all year. They share nesting duties as well, both birds taking an active part in building the nest, incubating and brooding.

Masked Owl
Tyto novaehollandiae

Other Names: Cave Owl, Chestnut-faced Owl

Distribution: Coastal regions of Australia generally, but not seen beyond 300 kilometres off the coast.

Notes: The Masked Owl can be found in heavily wooded eucalypt forests around the coast of Australia. By day it roosts in big hollows in trees, crevices in cliffs, and caves, but rarely in heavy foliage. At night it hunts for food: small mammals, reptiles and occasionally birds. Prey is mostly killed and taken back to the owl's perch to be eaten. If, however, the kill is too heavy to carry the bird will tear it into manageable strips and eat it on the spot.

  Masked Owls mate for life, and hold the same territory all year round. The female Masked Owl is about 10cm larger than her mate. They breed at anytime, and the approach of breeding brings extended periods of twittering from both birds around the nest hollow. The nest they construct is a deep, vertical hollow in a tall tree, or on a ledge in a cave. To enter the nest, the birds slide down tail first. This is a characteristic of the whole Masked Owl group.

Barn Owl
Tyto alba

Other Names: Delicate Owl, Screech Owl, White Owl, Lesser Masked Owl

Distribution: Found in most types of open wooded country but not dense forests, throughout Australia.

Notes: The name Barn Owl originated in Britain where this bird is often found roosting in barns because large numbers of mice and rats also live there. The Barn Owl, like all members of the Tytonidae family has a distinctive heart-shaped disc of facial feathers. It is found throughout the world in slightly varying forms.

The Barn Owl lives alone or in pairs, roosting by day in tree hollows, thick foliage, caves, rock crevices and buildings. At night they hunt mice, small birds, reptiles and night flying insects. They hunt usually by flying on steady, silent, beating wings and also by perching low to the ground. It appears that Barn Owls hunt more by ear than by sight. Regurgitated pellets of indigestible parts of prey are found beneath their daytime roosts, and these tend to create a smell.

Barn Owls are nomadic birds, their territory being determined by the availability of food. The numbers of owls also is proportional to the food supply. In a good season they breed prolifically; in a poor season many birds die of starvation.

Grass Owl
Tyto longimembris

Other Names: Eastern Grass Owl

Distribution: Coastal eastern Australia from Cape York to around Harrington, New South Wales, and from the Gulf country south through western Queensland into north-eastern South Australia.

Notes: The Grass Owl is a rarely seen bird. Its habitat is limited to the coastal heath and floodplain grasslands across northern Australia where the population seems stable and permanent. The owl rests during the day in its "hide" on a platform of heavily trampled grass beneath the undergrowth. At dusk it leaves the "hide" in search of food. The owls fly above the ground on silent, beating wings in search of prey which includes large insects, reptiles, frogs, and mostly, canefield rats. When they have sighted a target they dive on it and secure it with their talons. They eat their kill immediately by removing the entrails, snipping off and swallowing its head, and lastly consuming the body in manageable pieces.

The Grass Owl population in the interior is particularly prone to fluctuations in its food supply. In plague years the owls thrive as the Long-haired Rats, on which they live, are everywhere. When the plague finishes, however, the Grass Owl's numbers also drop, and they either die or have to find food elsewhere, often travelling as far south as Melbourne to find it.

Powerful Owl
Ninox strenua

Other Names: Great Scrub owl, Eagle-owl

Distribution: South-eastern Queensland to eastern and southern Victoria.

Notes: The Powerful Owl is the largest of Australia's owls. It inhabits wet sclerophyll forests and scrubs, particularly along the Great Dividing Ranges. During the day it roosts in trees. It appears to have a number of favoured roosts which it visits in regular rotation. Sometimes it will be found on one of these clasping a kill from the night before, and it will hold it all day in its talons before eating it in the evening.

The Powerful Owl can be distinguished by its mournful "woo-hoo" which can be heard at anytime of the year, mostly at dusk and at dawn. At night it hunts for food. The Powerful Owl kills medium sized tree-living mammals like the Great Glider, the Ringtail Possum and the Sugar Glider. It will also eat birds and young rabbits. Like most owls it tears its prey apart and consumes the head first.

Southern Boobook
Ninox novaeseelandiae

Other Names: Spotted Owl, Boobook Owl, Mopoke, Red Boobook Owl, Morepork, Marbled Owl, Cuckoo Owl, Fawn-bellied Owl, Tasmanian Spotted Owl

Distribution: Throughout Australia, including Tasmania.

Notes: It is still being debated as to whether the Southern Boobook and the New Zealand Boobook are two distinct species or merely two races of the one species of owl. The resemblance is very strong and they differ only in superficial ways, so it is very difficult to tell them apart.

The Southern Boobook is the smallest and most common of all Australian owls. They are widespread in wooded areas generally, from coastal rainforests to sparsely timbered regions of the interior. It is not often seen, however, but is more likely to be heard calling its "mo-poke" or "more-pork" from a perch somewhere. If the bird is disturbed by humans it will sit bolt upright and press its feathers tight to its body. Then it will turn sideways and its appearance becomes that of a very long and slender bird. In this position it is difficult to see, however its presence may be revealed by a mob of smaller birds harassing it until it is sometimes forced to take flight.

Boobooks prey on small mammals, night flying insects and moths. They can sometimes be seen flying around street lights in search of large moths.

Superb Lyrebird
Menura novaehollandiae

Other Names: Lyretail, Native Pheasant

Distribution: Wet eucalypt and rainforests of coastal eastern Australia and Great Dividing Range south to Melbourne, Victoria and north to Stenhope, Queensland; introduced to Tasmania.

Notes: Probably the most striking feature of the Superb Lyrebird is that it is a remarkably competent mimic, imitating almost any sound, but generally mimicking the calls of other birds and mammals in its habitat. Both males and females will imitate, but the female tends to be less vocal.

The other striking feature of the male is its display during courtship. He builds a number of display mounds, and visits them in turn stopping to sing and display. He spreads his tail like an opened fan over his back and head, showing the lyrate feathers and silvery filamentary plumes which give the bird its name. The female is attracted by the displaying male and will mate with him without forming a bond. The male, in this way, often mates with more than one female.

Nest building, incubating and rearing the young are the job of the female lyrebird. Their favourite positions for nest building are in banks, tree-ferns, among rocks or even in the forks of trees, and they use sticks, twigs, dried fern and mosses to build them.

Malleefowl
Leipoa ocellata

Other Names: Lowan, Gnow, Mallee Hen

Distribution: Throughout dry inland of southern Australia, mostly in mallee and other dry scrubs in semi-arid zone.

Notes: The Malleefowl is an intensely interesting bird, however, it has become reduced in numbers due largely to the clearing of its habitat for crops and grazing. This quiet, elusive bird is difficult to find; rather an observer is more likely to find the mound where Malleefowl eggs are incubating.

This mound is built by both the male and female, who mate for life, but still tend to spend their lives apart. Through a process of clearing and then rebuilding, they shape the mound until it is about 3-4 metres in diameter. An egg chamber is then dug in the top and filled with vegetable debris which is left until wet from rainfall. The mound is then totally covered in sand until it is about 1.5 metres high.

When the female is ready to lay the eggs, the male will dig holes in the warm, rotting vegetable matter in which she deposits them. The male then maintains the nest, checking that the internal temperature remains constant at 33°C. When the chicks hatch they must dig their way through about a metre of loose sand to the surface. From there it is completely on its own, the parents job having been completed, and within 24 hours the chick can fly.

Brush Turkey
Alectura lathami

Other Names: Scrub Turkey

Distribution: Eastern Australia from Cape York south to the Manning River.

Notes: The Scrub Turkey inhabits rainforests near the coast and scrubs of the interior. Although it can fly it spends most of the time on the ground, feeding on insects, native fruits and seeds raked up from the forest floor.

During the breeding season the male Scrub Turkey tends to the mound built in the previous year. Only the male builds the mound by grasping leaf material in its feet and flinging it backwards on to the site. As the male scratches away all the available leaf matter, he covers the mound with dirt.

Eggs are laid at intervals of two or three days. The female digs a hole in the top of the mound, lays the eggs and then covers the hole. Although she can lay around 14 - 18 eggs, many are eaten by goannas and large snakes. It appears that the number of eggs she lays is affected by the rainfall. The more rain which falls, the larger her clutch of eggs will be.

Blue-winged Kookaburra
Dacelo leachii

Other Names: Leach's Kookaburra, Howling Jackass, Barking Jackass

Distribution: From Shark Bay, Western Australia through northern tropical Australia to south-eastern Queensland; also in New Guinea.

Notes: The Blue-winged Kookaburra is the closest relative to the Laughing Kookaburra. It prefers, however, a wetter habitat than its relative and is usually found in eucalypt and gallery woodlands along creeks or in swampy coastal paperbark, thus is seen largely along the northern coastline of Australia.

While both species of Kookaburra have similar feeding, mating and nesting habits, there are two marked differences between them. As its name implies, the Blue-winged Kookaburra is distinguished by its blue rump, shoulders and flight feathers, while its call is more of a frantic, raucous, high-pitched, trilling howl.

The Kookaburras form permanent pairs and spend a long time in rearing their young. They nest in hollow tree cavities, chambers tunnelled into tree bound termites' nests and in baobab trees. The female lays one to four eggs. After reaching maturity, the young stay with their parents, helping to defend their territory and helping raise further offspring. The young birds may remain in this auxiliary role for four or five years and appear to form about one-third of the adult population. This is part of the kookaburra's form of "birth control".

Laughing Kookaburra
Dacelo novaeguineae

Other Names: Giant Kingfisher, Great Brown Kingfisher, Laughing Jackass, Settler's or Bushman's Clock

Distribution: In Eucalypt woodlands and open forests of eastern Australia from Cape York Peninsula, south throughout eastern Australia to South Australia; introduced to Western Australia and Tasmania.

Notes: The kookaburra is the world's largest kingfisher, but unlike its relatives, it is a sedentary bird. It is best known, however, for its laughing call, strongest in the early morning and at sunset. A calling Kookaburra is unmistakable as it squats on its haunches with its tail raised and head thrown back. They are a fairly common site even in city areas as they can be tamed quite readily by people willing to feed them.

In their woodland and forest habitats they feed on small reptiles, insects, crabs and fish. They hunt through the perch and pounce method. The bird will patiently sit motionless on a vantage point about 10 metres high watching the ground below intently. When they sight suitable prey they flutter down, seize it in their bill and then return to their perch to eat it.

A Kookaburra may live for 20 years or more. As a result their birthrate is low and the population turnover very slow. They tend to live in small groups which jointly maintain a territory and which co-operate in raising the young.

Rufous Whistler
Pachycephala rufiventris

Other Names: Rufous-breasted Whistler, Rufous-breasted Thickhead, Echony, Mock Whipbird, Thunderbird.

Distribution: All types of open forest and woodland throughout mainland Australia.

Notes: The Rufous Whistler gets its name from the male of the species who has a browny-red breast, while the female is much greyer in colour. It is a commonly found bird and there is hardly an area of open forest anywhere in Australia without it.

During the breeding season, pairs of Rufous Whistlers establish territories. This is all done initially by song until it builds into chasing flights between two males. Between the courting couple there is much displaying as the birds sing in chorus, and then one will adopt a begging posture as it follows its mate. The male goes in search of suitable nest sites but the decision is made by the female who proceeds to build the nest. Only the female incubates the nest at night but both brood and feed the young.

During the day the Whistler is to be found amidst the branches methodically searching for insects and occasionally berries and fruits. Sometimes they forage on the ground but most feeding is done in the trees.

Australian Magpie
Gymnorhina tibicen

Other Names: Black-backed Magpie, White-backed Magpie, Western Magpie, Long-billed Magpie, Flutebird, Piping Crow-shrike

Distribution: Eucalypt woodlands throughout most of Australia.

Notes: The Magpie is one of the best known birds in Australia. Most people are either likely to have been swooped by a protective magpie during their breeding season, or have heard it carolling its characteristic song at dawn or dusk.

The Magpie is a sedentary bird living in territorial groups with a defined social structure. The breeding season brings the magpies vying for social position and territory. The dominant male in a group may mate with any or all of the females in his group. The female then nests on her own, building and incubating without help. The dominant male usually only feeds one of the brooding females, thus it is rare for any other undefended and unsupported nests to produce chicks.

The magpie leaves its nest every morning to fly to fields where it eats. It feeds on insects and small snakes, walking methodically about, looking with its head on its side and jabbing its bill into the ground. In the evening it flies back to its roosting tree and sings duets as the sun sets.

Sacred Kingfisher
Halcyon sancta

Other Names: Green Kingfisher

Distribution: Australia generally, and Tasmania.

Notes: A breeding pair of Sacred Kingfishers do everything together. They begin by excavating their "burrow" 1 - 20 metres above the ground in an arboreal termite mound or hollow limb. They fly at the chosen site like a guided missile and make an indentation to mark the spot. They then dig using their bills as picks, and their feet to scratch away the dirt. Later the pair share incubation, rearing and defending together.

The Sacred Kingfisher is the most familiar of the smaller Australian Kingfishers with its beautiful turquoise blue and green colourings. It can be found in tall, open eucalypt forests and woodlands, paperbark forests, mangroves and sometimes along wooded rivers. Their diet consists of small reptiles, crickets, grasshoppers, beetles and their larvae, fish and crustaceans.

Rainbow Bee-eater
Merops ornatus

Other Names: Australian Bee-eater, Rainbow-bird, Pintailed Bee-eater, Gold-digger, Golden Swallow, Goldminer, Spinetail.

Distribution: Australia generally, throughout drier Australian woodlands, wherever suitable habitats are found.

Notes: Although there are 24 species of Bee-eater, the Rainbow Bee-eater is the only one found in Australia. It arrives in southern Australia in September or October and departs in February or March where it winters in the far north of the continent and on islands north of Australia. They are communal birds and live in groups of 20 to 30, all roosting together at night in the same tree.

They eat large quantities of bees, but also consume wasps, dragonflies, damsel flies, beetles, termites, locusts and moths. All food is taken in flight and then the Bee-eater returns to a perch to squeeze out the sting before eating it.

Rainbow Bee-eaters lay their eggs in nesting chambers at the ends of metre long tunnels dug in sandy soil. They dig together, loosening the soil with their bills and then use their legs to kick out loose dirt. Young chicks are prone to be taken by predators such as snakes and goannas who hear their calls in the nesting chambers.

Black-faced Cuckoo-shrike

Coracina novaehollandiae

Other Names: Blue Jay, Grey Jay, Shufflewing, Summerbird, Cherry Hawk

Distribution: Throughout Australia, including Tasmania.

Notes: The Black-faced Cuckoo-shrike is the most common of the seven species of cuckoo-shrike found in Australia. It inhabits open forests and woodlands and may be seen in city and suburban parks and gardens, as well as in the remote interior. Although similar to the cuckoo in form and flight, and possessing shrike-like bills, they are not related to either of these families of birds.

They are a partly nomadic, partly migratory bird, who, in late summer and autumn, roam widely in loose flocks, generally in a northerly direction. The Black-faced are tree living birds which feed on insects and their larvae and, sometimes, berries. They take their prey by pounce-diving on it from an open vantage point. They do this because they have strong wings but rather weak feet. Occasionally they will hover over the ground looking for food.

The Black-faced Cuckoo-shrike builds a tiny nest in the forked branch of a tree up to 10 metres from the ground. Their nests are very small in proportion to the medium sized bird and very soon the chicks become too small to sit comfortably in the nest. When this happens the chicks have to sit on top of the nest and, as a result, are sometimes blown away.

Willie Wagtail

Rhipidura leucophrys

Other Names: Shepherd's Companion, Black and White Fantail, Australian Nightingale, Water Wagtail, Morningbird

Distribution: Throughout Australia; accidental to Tasmania.

Notes: The Willie Wagtail is one of the best known and best loved of our birds and can often be found in gardens, school grounds, city parks, scrublands, open forests and grazing land. Any cleared land with a few exposed stumps or posts is ideal for wagtails, and where none of these are to be found the birds will perch on the backs of cattle and horses. From here they will launch out to snap up insects to eat.

The Willie Wagtail has a distinctive song which it sings, particularly at night. The chief call resembles the phrase 'sweet pretty creature' which it utters frequently. When alarmed, the bird gives a series of calls much like the sound of a rattling, half-empty match box. The bird also has a distinctive white eyebrow which is used in an aggressive display when two males meet during the breeding season. The eyebrow can be widened to show anger or narrowed to indicate submission. In this way the birds can 'fight' without any physical contact.

The hunting wagtail covers the ground in a series of hops and low zigzagging flights. Before take off and after landing, they flash their wings and fan their tails. Even at rest the bird sways its body and wags its tail from side to side, hence its name of wagtail.

Common Bronzewing
Phaps chalcoptera

Other Names: Forest Bronzewing, Bronzewing

Distribution: Throughout Australia and Tasmania, where suitable habitats exist, except for most of Cape York Peninsula.

Notes: The Common Bronzewing occurs in almost any kind of wooded country but avoids dense rainforests. They occur in greatest numbers in sub-coastal and inland woodlands especially where there is mulga, or other acacias. They feed on the seeds of grasses and herbaceous plants, native fruits and berries. Much of the Bronzewing's range is now wheat country, and in this area the bird will eat waste wheat grains.

Common Bronzewings usually live in pairs and together they roost and nest in trees. They spend their days on the ground under a bush where they feed and call their mournful 'oom' to each other. They forage actively in the early morning and late afternoon and drink before dawn and immediately after dark. The Bronzewing can usually be found on the ground preferring to walk from place to place. When flushed, it rises with a loud whirring noise, flies rapidly for a short distance and then alights on a tree branch.

Scarlet Robin
Petroica multicolor

Other Names: Robin Redbreast, Scarlet-breasted Robin, White-capped Robin

Distribution: From the Darling Downs in Queensland, through New South Wales to Victoria, and across to South Australia; also in Tasmania. An isolated race occurs in Western Australia.

Notes: The Scarlet Robin is basically a bird of the open forest but in summer it moves to dense forests at low altitudes and can often be found in the hills surrounding Canberra. They are very like their family member the Flame Robin and have similar habits but are quite distinguishable with their prominent white caps.

The robins move quietly around the forest in pairs foraging for food. Their favourite hunting technique is to perch on a low vantage point and sit statue still, except for the occasional flick of the tail. They will scan the ground below and then when they spot their prey they will flit suddenly to the ground and seize it. Then they will return once again to their watch post. Scarlet Robins feed on grasses, small insects, grubs and worms.

The female robin weaves an open cup shaped nest from strips of bark, mosses and grasses, and will line it with wool, feathers, hair or fur. Usually it is placed in the fork of the tree and can be quite low to the ground. Here a mating pair of birds will raise up to three broods of chicks in one season. The male sings a territorial ringing song and both birds will trill a sweet song of nine notes during the day.

Flame Robin

Petroica phoenicea

Other Names: Robin Redbreast, Flame-breasted Robin

Distribution: From south eastern Queensland through south eastern Australia to south eastern South Australia, Kangaroo Island, the islands of Bass Strait, and Tasmania.

Notes: It is not an uncommon sight to see a Flame Robin perched on a fence post or low branch flicking its wings and tail, watching the ground for insects. It will then take flight appearing to play-chase by flying low and swooping to the ground, and then darting off again. The Flame Robin is migratory, frequenting farmlands, golf-courses, parks and open woodlands in the colder months and then withdrawing to the highlands in the Spring.

During the breeding season they form territory-holding pairs. While the female builds her cup-shaped nest the male sings to advertise their territory. After breeding, the robins form loose family groups of 5-30 or more, moving after summer from their breeding grounds into lower open country. This species is very common in Tasmania and it may migrate between that island and Victoria.

Mistletoebird
Dicaeum hirundinaceum

Other Names: Australian Flowerpecker

Distribution: Throughout mainland Australia, except Tasmania.

Notes: The tiny Mistletoebird gets its name from the mistletoes on which it almost exclusively feeds. In fact the bird and the plant are dependent on each other for their continued survival. The bird eats the berries of the mistletoe, and defecates the seeds within 60 minutes of ingesting them. The seeds are very sticky and glutinous, and will adhere to any surface. These seeds then germinate and if they have lodged on a suitable host tree, a mistletoe plant will establish.

The bird depends so greatly on mistletoe fruit because its digestive system has adapted to this specialised diet making it very difficult for the bird to digest little else than a few insects, and the berries. The birds are locally nomadic, moving from area to area as different strains of mistletoe tree come to fruition.

The Mistletoebird builds a soft, pear-shaped nest of plant down and spiderweb where it will raise three chicks. The chicks are fed for the first few days on insects but are quickly weaned onto mistletoe fruits.

Yellow-bellied Sunbird
Nectarinia jugularis

Other Names: Olive-backed Sunbird, Yellow-breasted Sunbird

Distribution: Rainforest edges, mangroves and suburban gardens in coastal Northern Queensland from Cape York to about Gladstone.

Notes: This small bird is a common sight in the tropics as it lives as much in suburban gardens as in the rainforests and mangroves. Wherever there are nectar-bearing flowers, these birds will be found.

The Yellow-bellied Sunbird is very much like the honey eater with its long curved bill and brush-tipped tongue for taking nectar. The Sunbird uses a variety of techniques for gathering the nectar from flowers. In front of small flowers it has the humming-bird habit of hovering and inserting its bill. With large open flowers they cling to the front, tearing the petals to get at the nectar. To get the nectar out of long, narrow trumpet-shaped flowers they pierce the base of the corolla.

The female Sunbird builds a long nest with a hooded side entrance. Often the birds will choose a site close to human habitation to build their nests and will use the same one year after year. Popular sites for building are under eaves, powerlines or verandas. Here the female incubates her eggs while the male attends her.

Striated Pardalote
Pardalotus striatus

Other Names: Eastern Striated Pardalote, Black-headed Pardalote, Yellow-tipped Pardalote, Chip Chip, Red-tipped Pardalote, Striated Diamondbird

Distribution: Almost the whole of Australia and Tasmania.

Notes: Until recently there were several recognised species of pardalotes with striped crowns, but now all of these are regarded as one species. The Striated Pardalote is a nomadic bird and is widespread across the continent. Its habitat is open eucalypt forests and woodgums, river-gum lined waterways and inland areas. Populations of the bird in the south often travel northward in autumn or winter, or leave high ground for lower altitudes.

Striated Pardalotes feed on the blossoms and outer foliage of eucalypts. Here they eat lerps, bugs, grasshoppers, beetles, cockroaches, thrips, weevils, ants, bees, wasps, flies and caterpillars. As they feed they call their distinctive "chip chip" to each other.

During the breeding season a mating pair of Striated Pardalotes will build a cup-shaped, partly domed, nest of grass, barkfibre and rootlets which they sometimes line with feathers. This they build in a hole in a tree stump, in a burrow in the ground, or in a bank.

White-plumed Honeyeater
Lichenostomus penicillatus

Other Names: Greenie, Native Canary, Chickowee

Distribution: Close and open woodlands over eastern mainland and across the inland on eucalypt-lined watercourses.

Notes: The White-plumed Honeyeater is one of the most widespread of honeyeaters and is known to frequent urban parks and gardens. It is a noisy bird with its squeaky "chip" which it sounds continually while feeding, to keep it in touch with others in its small flock. It is rarely still whether it is feeding or resting and can be seen either darting and bouncing around trees, or twisting and reversing when stationary.

The White-plume has a short beak which it uses to pick up manna and honeydew as well as the occasional beetle, bee, weevil, aphid or other insect it can find on eucalypt foliage. As in all honeyeaters, the tongue is muscled so that it can protrude far from the bill, and it is topped with over 50 fine bristles that soak up sugary liquids. The bird dips its bill into flowers and licks up the nectar at about 10 licks per second.

When breeding, females build and incubate while males attend and make frequent song-flights. The birds are aggressive in the defence of their territory and will send shrill alarm notes to each other in warning. A group of birds will often mob an intruder larger than themselves in an attempt to defend their territory.

Bell Miner
Manorina melanophrys

Other Names: Bellbird, Bell Mynah

Distribution: Coastal and mountain areas of eastern Australia from Mary Valley, Queensland to south western Victoria.

Notes: The clear, bell-like notes of the bellbird are some of the most beautiful sounds to be heard in the Australian bush. In gullies of the wet and dry sclerophyll forests where the birds live, the repetitive single note call is the bird's best way of keeping in touch with other members of the flock. Bell Miners live in colonies of about 10 birds which comprise a mated pair and additional members. These colonies are sedentary remaining for years in the same territory. Their feeding ground is the middle strata of the forest and here they feed on leaf-insects and occasionally nectar.

Breeding takes place throughout the year but mostly in spring or early summer. The female builds the nest, incubates and broods the young while the male escorts her. Many members of the group take part in feeding the young, often queuing up to do so, and they will also help clean the nest. The colony members also play an important role in defending their territory from other species and often perform strange displays in order to distract an enemy and lead it away from the breeding area.

Western Spinebill
Acanthorhynchus superciliosus

Other Names: None

Distribution: South-western Australia, extending north and east to Moora, the Stirling Range and Israelite Bay.

Notes: Spinebills are small, active honeyeaters with long, tapering bills for probing tubular flowers for nectar. They also have almost tubular tongues with short serrations, suited to their feeding forages in the well-shrubbed woodlands and heaths of south-western Australia. While nectar is the bird's staple diet, it also eats insects caught on the wing, or gleaned from shrubbery. Western Spinebills visit flowering shrubs as they come into flower — dryandras, banksias and various myrtles. As they feed they are brushed with pollen from the flowers, which is passed on at the next flower, and so on. In this way, pollination is effected.

These birds rarely move far with the season as there is always a shrub in flower. They breed from September to January. The female builds a small cup nest of bark fibre, fine stems and binding cobweb which she lines with fibre and banksia fur.

Superb Fairy Wren
Malurus cyaneus

Other Names: Blue Wren, Superb Warbler, Bluecap, Blue Bonnet, Cocktail, Fairy Wren, Mormon Wren

Distribution: Highland south-eastern Queensland to Tasmania and southern Eyre Peninsula, South Australia; also much of the Riverina.

Notes: The Superb Fairy Wren lives in thickets in the fringes of scrubs and the banks of water courses. It has also adapted to urbanisation, parks and gardens making it one of Australia's most popular and best-known birds.

The Superb Fairy Wren is a sedentary bird remaining paired in a close-knit group year round. They have a pretty, reeling song which they sound in the morning before daylight and then shortly after dawn. They spend the morning hopping and flitting along, pecking for food. During the middle of the day they rest and preen each other. Later in the afternoon they feed again before huddling together side by side to sleep in the evening.

There always appear to be a large number of female wrens in a group. This is usually not the case though, because immature males have the same brown colouring as the females. They moult before the breeding season into the blue "nuptial dress" which distinguishes the male from the female.

With the help of auxiliary members, pairs of breeding Fairy Wrens regularly raise two or three broods a season. Despite the large number of chicks born each year fewer than half will reach independence due to misadventure and predation.

Tawny-crowned Honeyeater
Phylidonyris melanops

Other Names: Fulvous-fronted Honeyeater

Distribution: South-eastern Australia from about Grafton, New South Wales to Tasmania and west to the Eyre Peninsula and Kangaroo Island, South Australia; also Western Australia from Israelite Bay to the Murchison River.

Notes: Tawny-crowned Honeyeaters inhabit dry, open heath and dwarf scrub. Here they feed mainly on nectar in low shrubs of epacrids, proteads, bottlebrushes and myrtles. They do a lot of feeding on the ground or very low to the ground in bushes. Often they will eat insects they have spotted while sitting watching. They will then fly out and pluck the insect. Occasionally they will enter forests to feed on nectar from blossoming eucalypts or to probe the bark for honeydew excreted by insects.

The male Tawny-crowned Honeyeater indulges in quite spectacular displays in spring in which he marks his breeding and feeding territories. He will flutter almost vertically up, then spiral back down on extended wings and tail, like a falling leaf. All the time he will sing his simple, melodious, wistful song. At other times he will perch on the top of a shrub and sing in short bursts.

New Holland Honeyeater
Phylidonyris novaehollandiae

Other Names: Yellow-winged Honeyeater, White-bearded Honeyeater, White-eyed Honeyeater

Distribution: Eastern and southern Australia from about Gympie, Queensland south to Tasmania and west to the Eyre Peninsula, South Australia; also in Western Australia from about Moora to Israelite Bay.

Notes: New Holland Honeyeaters are found in a variety of habitats including coastal heathlands and gardens with native shrubs, particularly banksias. They feed almost exclusively on nectar taken from blossom in shrubs and small trees. Research has shown that they also eat the sweet encrustations from damaged leaves, called manna, and the excretions of aphids, known as honeydew. Occasionally the birds will eat insects to supply extra protein, however, their tiny, thin-walled stomachs cannot digest much hard food, so they do not eat a great number.

New Holland Honeyeaters will drink from over 100 species of plants making them important in effecting pollination. As they force their bills and faces into the mouths of flowers, they become dusted with pollen. As they move on they transfer this to the next flower, and then the next plant, and so on. The birds spend three to ten hours a day eating in clearly designated feeding territories which they mark with song-flights each morning and afternoon. Often they will feed with other honeyeaters who will respond to the New Holland's warning calls of an intruder.

Pairs of New Holland Honeyeaters often nest in a loose communal group. Both male and female feed the young. Sometimes the male takes over care of the fledglings completely after they are about three weeks old, leaving the female to build and nest again. Sometimes a female may have two or three broods in one season.

Little Wattlebird
Anthochaera chrysoptera

Other Names: Western Wattlebird, Lunulated Wattlebird, Mock Wattlebird

Distribution: Heaths and shrubby woodlands and forest with flowering proteads in south western mainland.

Notes: The Little Wattlebird and the Brush Wattlebird have long been treated as the one species. The Brush Wattlebird inhabits shrubby forest and tall heath on coasts in south eastern Australia and Tasmania, but differs from the Little Wattlebird in many little ways that add up to some major differences in colour pattern and biology.

The Little Wattlebird feeds on nectar and pollen from native flowers such as banksias, eucalypts, grevilleas and mistletoes, and also on insects. They are often found in urban gardens tucked up in some native tree feeding. They are active, noisy birds — the females with high pitched calls which they often use when singing duets with the males. They breed mainly in July to November and build a small rough saucer nest of interwoven stems and twigs which they line with fine grass and down. Unlike the Brush Wattlebird which lays two or three eggs, the Little Wattlebird lays only one egg.

Zebra Finch

Poephila guttata

Other Names: Chestnut-eared Finch

Distribution: Australia generally, except Tasmania and high-rainfall areas of the far north, south-east and south-west.

Notes: Zebra Finches are common in all habitats except wet, coastal forests, and have cleverly adapted in various ways to their arid environment. Their metabolic rate is lower than other finches which means they excrete less water and can drink salty water not tolerated by other birds. They do, however, still need to drink every hour or two and so will always be found close to a watering hole. They drink by fully immersing their faces in water and sucking. The finches feed on the ground pecking up fallen seeds, mostly grasses. They call to each other with a soft trumpet like note as they fly to their evening roosts.

Zebra Finches frequent large flocks of up to 100 drinking, feeding, bathing and preening together, and will stay this way even through the breeding period. Because such large numbers inhabit the same area, more than one nesting pair will often be found in the same bush. Nest building is done after a site is chosen, with the male collecting the materials and the female constructing it. Their nest is flask shaped with a side entrance tunnel.

Blue-faced Honeyeater
Entomyzon cyanotis

Other Names: Banana-bird, Pandamus-bird

Distribution: Eucalypt, paperbark and pandamus woodlands to edges of rainforests and mangroves around north and east from the Kimberleys, Western Australia to south east South Australia.

Notes: The Blue-faced is one of the larger honeyeaters inhabiting woodland areas. Although their diet consists mainly of leaf beetles and weevils, which they forage for by probing and prising under paper bark, they have been known to drink the nectar from flowering paperbarks and grevilleas, and also to eat bananas and pears.

The Blue-faced Honeyeater is a communal bird. It lives in small groups of between 2-10 birds which eat and sleep together. At dawn and dusk, individual birds in the flocks sing strong, strident piping notes from the tops of trees. During the early morning and late afternoon they feed close together in the upper foliage of trees calling softly to one another. When feeding is complete they will often bathe and drink, diving down from their perches to splash in the tops of pools of water.

The Blue-faced Honeyeater does not always build its own nest, preferring instead to occupy deserted roosting nests of Grey-crowed Babblers. The Blue-faced rebuilds the lining at the top and then lays its eggs ready to hatch.

Dusky Woodswallow
Artamus cyanopterus

Other Names: Jacky Martin, Blue Martin, Bluey, Skimmer, Bee Bird, Sordid Woodswallow, Woodswallow

Distribution: Eastern Australia from near Cairns, Queensland south to Tasmania and west to Kangaroo Island and Eyre Peninsula South Australia; also extreme south-western Australia.

Notes: Dusky Woodswallows occur around coastal eastern and southern Australia in dense eucalypt woodlands and forests though they also frequent open spaces, partly cleared lands, orchards and parks. Like other Woodswallows, the Dusky flies gracefully over the tops of trees, gliding, soaring and fluttering.

The Dusky lives in small flocks of 10 to 30 birds. In winter these flocks group together to form a large flock of several hundred. En masse they huddle into tree chimneys to sleep in order to conserve body heat and energy. Together a flock of birds will defend against a predator, chattering amongst themselves to ward off the intruder.

The Dusky Woodswallow migrates northward in Autumn and southward again in Spring for breeding. The courtship routine involves a pair of birds on the same branch. The male offers the female food and she begs for it by extending her quivering wings, thrusting her head forward and opening her bill. Following mating a silent ritual of wing fluttering and fanning of the tail occurs between the mating pair.

Australian Hobby
Falco longipennis

Other Names: Little Falcon, White-fronted Falcon

Distribution: Throughout Australia

Notes: The Australian Hobby is to be found throughout Australia where there are trees, particularly in open wooded country. There are actually two races: one black-backed race is to be found in the coastal southeast and the extreme southwest of the mainland, and in Tasmania; the other race, the grey-backed is found everywhere else.

This fierce hunter preys mainly on birds, capturing them in mid-air either from direct pursuit or in dives from above. It will take birds up to its own size but has been known to stun a larger bird, knocking it to the ground. There it will pluck it and eat it, or lift it in its talons and carry it to a perch. The hobby prefers high, exposed branches on which to rest or search for movement of prey. It will also take to the wing to hunt, sometimes darting at tree-top level to surprise small birds below. Sometimes at night it will be seen feeding on bats and large flying insects.

The Australian Hobby breeds from September to November when a pair will occupy the old nest of a magpie or crow, or other bird, high in the trees. They line the nest with leaves or bark and will defend it aggressively and noisily. Here the female lays 2-3 eggs and incubates and broods them unaided while the male hunts. He calls the female from the nest to take his catches and passes them to her in mid-air or in a nearby tree.

Brown Falcon
Falco berigora

Other Names: Brown Hawk, Cackling Hawk

Distribution: Australia generally including Tasmania.

Notes: The Brown Falcon inhabits open and lightly timbered country and can often be seen flying over open spaces or perched on a telephone post or fence. It is well adapted to agricultural areas and towns and has been known to become rather tame in inhabited areas.

The Brown Falcon resembles the Black Falcon in appearance but is very different from any other falcon in its flight and hunting methods. It rarely hunts by chasing its prey on the wing. Instead it will watch from a high vantage point and then pounce on its prey, grabbing it with its talons. It feeds on snakes, grasshoppers, mice, and small birds. The wing beats and flight of this falcon are relatively slow. When it flies it alternates between beating its wings and gliding with them held up in a shallow V. At times it hovers, somewhat inefficiently, but it can also soar to incredible heights.

Brown Falcons are very variable in plumage. It has six main colour forms from dark to light brown or reddish-brown. Because of this variation, it has often been mistaken for more than one species.

Little Eagle
Hieraaetus morphnoides

Other Names: None

Distribution: Inland areas of Australia generally, extending to the coast in the west and north-west, and occasionally in eastern Australia.

Notes: In appearance the Little Eagle is very like a small Wedge-tailed Eagle, possibly because they are the only two eagles to have fully feathered "legs" down to their toes. They can usually be found in pairs in wooded country, especially where there are creeks. It is a sedentary bird spending most of its time high in slow wheeling flight on warm thermal currents of air, searching for prey.

The Little Eagle has a preference for live prey; rabbits, small mammals, reptiles, and occasionally other birds being its favourite. It tends to kill and eat its prey on the ground, though it sometimes carries its kill to a safe place where other scavenging birds will not see it. There they pluck and tear at the carcass with their talons and bill.

During courtship the male performs a spectacular display of flight, calling loudly to the female as he flies high, and then dropping steeply only to climb again.

White-bellied Sea-eagle
Haliaeetus leucogaster

Other Names: White-breasted Sea-eagle, White-bellied Fish-hawk

Distribution: Throughout coastal mainland Australia and in Tasmania; about some of the larger inland rivers and lakes.

Notes: The White-bellied Sea-eagle is an attractive bird with grey-white tonings that can often be observed flying slowly above foreshores, mud-flats, or sandspits, or circling and soaring high in the air on its broad upswept wings. They scavenge along beaches searching for offal and carrion left by the tide, or for prey such as tortoises, sea-snakes and waterfowl.

In the breeding season they display by soaring and calling to each other with their peculiar metallic, crackling cry. They catch fish and then soar high into the air with it, dropping it and diving to catch it again in midair. The female lays her eggs several days apart but incubation begins with the laying of the first thus giving the first hatched a head start. Because of this, the older chick often takes all the food leaving the younger ones to die.

Black-shouldered Kite
Elanus notatus

Other Names: None

Distribution: Australia generally, but not Tasmania.

Notes: The Black-shouldered Kite can be found throughout mainland Australia in woodlands and wetter savannas. It is the Australian version of a worldwide group of 'black-shouldered' kites varying from its relatives in only subtle ways.

The Black-shouldered Kite is a nomadic bird, moving from district to district according to food supplies. It is often seen in areas surrounding farms, fluttering and gliding on upswept wings searching for food. When it sights its prey it hovers on its tail, then silently drops down on to the victim, dangling its talons outstretched and with its wings swept up above its head. It eats rodents, amphibians, reptiles, and insects.

The courtship flying of this bird is interesting. They will soar and flutter together, but sometimes the male will dive at the female. She will turn in the air, the pair will lock claws and the female will carry the male whirling down.

Brahminy Kite
Haliastur indus

Other Names: Red-backed Sea-eagle, Rufous-backed Sea-eagle, White-headed Sea-eagle, Red-backed Kite

Distribution: Coastal northern and eastern Australia west to about Shark Bay, Western Australia, and south to Port Macquarie, New South Wales.

Notes: The Brahminy Kite is an attractive and easily recognised bird with its chestnut and white plumage. It inhabits the mangrove-lined coastal inlets and bays of northern Australia and can be seen around rocky shores and beaches. In Australia it is a solitary bird, and does not gather in flocks as it does in India. Rather it pairs only to breed, and at other times appears to patrol its own designated area of coastline alone.

The Brahminy Kite feeds on fish, crabs, reptiles, insects and occasionally small birds and rodents. It only rarely takes live prey, preferring, instead, fish and other marine creatures cast up by the tide. It is unable to kill large, live animals. Sometimes the kite takes its prey from the water or ground with its talons and carries it to a perch in a tree where it will kill and eat it. Generally, however, it prefers to eat its prey where it is caught.

During the breeding season the male and female will build a large nest, high in the fork of a tree, made of sticks which they line with fine bark, grass or leaves. Occasionally they will decorate the sides of the nest with streamers of bleached seaweed, The female then incubates the eggs while the male hunts and brings her food. Both parents hunt for food after the chicks hatch.

Sulphur Crested Cockatoo
Cacatua galerita

Other Names: White Cockatoo, Yellow-crested Cockatoo

Distribution: Northern and eastern Australia, west to the Kimberleys and south to Tasmania and south-eastern South Australia; also in New Guinea.

Notes: The Sulphur Crested Cockatoo is one of Australia's best known birds common to many towns and cities. Its habitats are mainly heavily-timbered mountain ranges, open forests, paddocks and the timbers bordering watercourses.

The cockatoo lives in a large flock, each flock having its own roosting tree. At night the birds return from their feeding ground to the roosting tree. Here they vie with each other for the best positions in the tree, screeching loudly and squabbling with each other. The noise often does not subside until well after dark, and it begins shortly after sunrise the next day.

Often the flock will have to travel to a suitable feeding tree. Here the majority of members in the flock will feed on the seeds, grains, berries or nuts in the trees or bushes, while sentinel birds will perch nearby keeping a look out. If they sound the warning of an intruder the whole flock will immediately rise into the air.

Gang Gang Cockatoo
Callocephalon fimbriatum

Other Names: Red-headed Cockatoo, Red-crowned Cockatoo, Helmeted Cockatoo

Distribution: Eastern Australia from Muswellbrook, New South Wales south to Victoria and King Island on Bass Strait; introduced to Kangaroo Island.

Notes: The Gang Gang lives chiefly in heavily timbered, mountain eucalypt forests in south eastern Australia as well as in the dense coastal forests of Victoria. They feed mainly on the seeds of native shrubs and trees, such as eucalypts, acacias and cypress pines by perching in the trees and cracking the seeds open with their strong bills. The birds will return each day to the same tree until the food there is exhausted. Then they move on to the next tree leaving behind them the remains of their feeding littered on the ground. The Gang Gang tends to become oblivious while it is eating and so it is possible to move close enough to almost touch them.

The cockatoos breed in the mountain forests. During the winter months they will move down into the coastal regions and can often be found in the gardens and parklands of Canberra and the outer suburbs of Melbourne. During this time they gather in small flocks seeking out food together. Sometimes a flock of up to 100 birds can be seen feeding in the one tree, and then suddenly they will leave the tree to fly overhead in wide circles, screeching loudly all the time.

Galah
Cacatua roseicapilla

Other Names: Goulie, Roseate Cockatoo, Rose-breasted Cockatoo, Willie-Wilcock.

Distribution: Continental Australia generally, chiefly inland areas.

Notes: The Galah is one of our more common species and may be found frequenting open country, especially inland plains, open grasslands and savanna woodlands.

They are usually seen in very large flocks — anything from 30 to 1000 birds — seeking food, or flying en masse. The birds in the air present a striking sight of colour as they tilt and turn, one moment pink, the next grey.

In the breeding season the birds pair up ready to mate, and search out a nest tree to which they will return year after year. A strange habit of the nesting Galahs is the male's habit of removing bark from a particular place on their nesting tree. Then, throughout the nesting season both birds will visit this stripped patch wiping their bills and faces over it, probably to mark their ownership of the tree.

Australian King Parrot
Alisterus scapularis

Other Names: Southern King Parrot, King Lory

Distribution: Common in heavily timbered mountain rainforests from Cooktown to southern Victoria.

Notes: These beautiful parrots are fairly common and conspicuous but they are very wary birds, taking flight when disturbed and calling loudly to each other with their shrill, grating "eek-eek-eek". Unfortunately the clearing of land and cutting of the trees has greatly diminished the habitat of the parrot, which is now protected by law.

King Parrots are to be found in heavily timbered mountain ranges feeding in the outer most branches of trees, particularly eucalypts and acacias. They feed on native seeds, fruits, berries, nuts and blossom, and in certain districts have been known to cause considerable damage to maize crops.

The King Parrot like some other species of bird is sexually dichromatic; that is different colours between the sexes. While the male is much more colourful with red, blue and green colouring, the female tends to be green all over.

Pink Cockatoo
Cacatua leadbeateri

Other Names: Major Mitchell Cockatoo, Leadbeater's Cockatoo, Wee Juggler, Cockalerina

Distribution: From north-western Australia through central Australia to south-western Queensland, western New South Wales, western Victoria and the inland parts of South Australia.

Notes: The Pink Cockatoo is probably best known to Australians as the Major Mitchell, named after Sir Thomas Mitchell who wrote about the bird on his journey through the interior of New South Wales. This popular bird, however, is under threat due to the loss of its natural habitat to development and through trapping for the illicit bird trade. They have difficulty recovering their numbers because they need several kilometres of clear, undisturbed land for breeding, and this is becoming restricted.

The Pink Cockatoo is usually found in pairs, or small groups, but rarely in very large flocks. They spend much of the day feeding on the ground or in the branches of trees and shrubs. Their diet consists of seeds, nuts, fruits and roots of acacias and cypress pines.

During courtship the male cockatoo struts along a branch towards the female with his crest raised. He chatters softly to her, bobbing his head up and down and moving it from side to side in a figure eight movement. The pair will then preen each other. When breeding, the birds become aggressively territorial and keep the breeding site free of any other birds. Both birds will incubate the eggs — the male doing the day shift, the female the night — and feed the chicks. Later the fledglings will remain with the parents to form a family group.

Red-winged Parrot
Aprosmictus erythropterus

Other Names: Crimson-winged Parrot, Red-winged Lory

Distribution: Northern and eastern Australia, west to the Kimberleys and south to northern New South Wales (generally west of the Great Dividing Range) and north-eastern South Australia.

Notes: The Red-winged Parrot is closely related to the Australian King Parrot, however it has its differences. They are not as communal as the King, preferring to live in pairs or small family groups, frequenting scrub and open forests, especially along water courses. They tend not to gather in feeding flocks, but feed separately on seeds, berries, nectar and insect larvae among the outer branches of trees. They come to the ground only to drink.

Red-winged Parrots have the unusual habit of scratching their heads by pushing their foot forward under not over the wing. When they fly their wings beat with deep irregular flaps, and their colouring makes them a very pretty sight in the air. Females and immature males differ from the adult male in being duller in colour. They also lack the dark blue back and rump the male has which he exposes during courtship.

Princess Parrot
Polytelis alexandrae

Other Names: Alexandra's Parrot, Rose-throated Parrot, Rose-throated Parakeet

Distribution: Far inland parts of central and northern Australia to the Northern Territory and northern South Australia.

Notes: The Princess Parrot's habitat is generally belts of timber near watercourses. Little is known about this bird who lives in the arid interior, however, it is known to be highly nomadic moving over long distances searching for seeds of various plants, particularly porcupine grass. They also eat acacia blossoms and some Mistletoe berries. It seems that a pair of birds, or a small flock, will appear at a particular site on a treelined watercourse, and will stay to breed. Then they will disappear as abruptly as they arrived perhaps not to return to that site for some 20 years.

Several pairs of Princess Parrots may form a small breeding colony. Up to ten nests have been found in the one tree to suggest this. When disturbed the parrots fly to the nearest tree where they tend to perch lengthwise along a stout limb, probably to help them avoid detection. They have some resemblance to another breed, the Superb Parrot, in flight and courtship display.

Yellow-tailed Black Cockatoo
Calyptorhynchus funereus

Other Names: Funereal Cockatoo, Yellow-eared Black Cockatoo, Wylah

Distribution: Coastal eastern Australia west to the Eyre Peninsula and Kangaroo Island, South Australia, and south to Tasmania.

Notes: The Yellow-tailed Black Cockatoo is to be found in pairs or flocks in heavily timbered mountain ranges, adjacent eucalypt forests and pine plantations, as well as in banksia scrub during winter. The bird has a slow laboured flight and while in the air can usually be heard calling with long-carrying, wailing cries.

There are two races of Yellow-tail. The race which inhabits Tasmania and western Victoria and South Australia appears to eat seeds exclusively, particularly enjoying radiata pine. The other race eats both seeds and insect larvae. Both races enjoy the seeds of banksias, casuarinas and hakeas. They have narrow and sharp-pointed bills for biting into the seed bearing cones from trees. The insect eating race also uses its powerful bill to make deep holes in a branch or trunk to retrieve insects and grubs.

Palm Cockatoo
Probosciger aterrimus

Other Names: Great Black Cockatoo, Goliath Cockatoo, Cape York Cockatoo

Distribution: Cape York Peninsula, south to the Rocky River.

Notes: The Palm Cockatoo is a beautiful, large black cockatoo — the only cockatoo in Australia without a coloured band on its tail — with a spectacular crest and crimson cheeks. It is confined to the tip of the Cape York Peninsula where it lives exclusively in the rainforest. It roosts singly among dead or leafless branches at the tops of tall trees on the edge of the forest.

In the morning they preen themselves by dusting their feathers with a powder from the base of the tail. This powder gives a slaty cast to their black plumage. Sometimes up to seven birds will congregate in the tree after preening, and will indulge in a variety of elaborate displays before flying out to seek food. The Palm Cockatoo eats seeds, nuts, leaf buds, fruit and berries, and grubs. The bird uses its extremely powerful jaw muscles to help it peel off or cut through seed coatings. This same bill, when flying, remains open and pressed to the breast because the cockatoo cannot close it due to the angle of the jaw.

Long-billed Corella
Cacatua tenuirostris

Other Names: Slender-billed Corella, Long-billed Cockatoo

Distribution: From Hamilton, Victoria to the Coorong, South Australia, ranging to Melbourne and southern Riverina, New South Wales.

Notes: In south-western Victoria a concentration of their food supply attracts large flocks of several hundred birds, giving the impression that Long-billed Corellas are plentiful. In fact these birds are common to their given localities but are rare anywhere else. It was believed the birds were threatened with extinction, however, it appears that the last two decades have seen an increase in their numbers.

The long-bills live in open forests, woodlands and grasslands bordering watercourses. They are noisy, conspicuous birds who indulge in pre-roosting acrobatics where they fly through the trees screeching loudly. During the day they spend most of their time feeding on the ground. They eat seeds, roots and bulbs and the constant digging for food leaves their underparts stained with dirt and vegetable matter. The long-bill uses its long, tapered bill to help it dig. They drink early in the morning and late at night.

The Long-billed Corella uses a sentinel warning system like that of a Sulphur Crested Cockatoo. While the main flock is feeding on the ground, a few birds remain in the trees above to look out for danger. On their warning of an intruder, the whole flock rises in to the air.

White-Cheeked Rosella
Platycercus eximius

Other Names: Eastern Rosella, Pale-headed Rosella, Mealy Rosella, Northern Rosella, Smutty Rosella, Brown's Rosella, Golden-mantled Rosella, Rosehill Parakeet.

Distribution: Eucalypt woodlands, open forests and their edges throughout eastern and northern Australia inland to fringes of inland plains; seven races.

Notes: This is the most common and best known of all Rosellas. It was originally called the "Rose Hiller" by the first settlers of New South Wales, getting its name from the farming area of Rose Hill in Parramatta, but over time the name became Rosella. There are three main races of White-cheeked Rosella, and because they differ quite markedly from each other they are often listed as separate species.

White-cheeked Rosellas inhabit open eucalypt country and partly cleared lands, and are seldom found above 1000 metres in the mountains. They are not a travelling bird, preferring to remain in the vicinity of their birth. Here, in small family groups, they feed together on the ground, hunting for seeds, fruit and insect larvae. They will eat in trees, especially fruit trees, where they devour the blossom, nectar and pulp of fruit.

The birds drink in the morning and evening, resting during the day in the tops of trees. They fly low through the trees, quick-fire flapping one minute, and swooping glides on folded wings the next. Then they glide upwards to alight on a branch with their tails fanned.

Crimson Rosella
Platycercus elegans

Other Names: Crimson Parrot, Pennant's Parakeet, Lowry, Mountain Lowry, Yellow Rosella, Adelaide Rosella, Murrumbidgee Lowry, Murray Smoker.

Distribution: Eastern Australia from the Atherton Tablelands, Queensland to Victoria and extreme south eastern South Australia.

Notes: There are three main races of Crimson Rosella: the red, the orange and the yellow. Their habitats vary from the rainforests of Northern Queensland, to the hill and mountain eucalypt forests of south eastern Australia, from the river red gums of the Murray, down to the sclerophyll forests of Kangaroo Island.

Crimson Rosellas are gregarious by nature. Young birds tend to move in flocks of up to 30 birds, wandering in search of the seeds from eucalypts, fruits, berries, to blossom and cyprus seeds. Adult birds, on the other hand are more likely to be found in smaller groups of five or six until the breeding season when they pair.

Pairing appears to be permanent. The male bird in courtship bobs up and down, chattering all the time while he displays his fluffed up feathers and spread tail. The female incubates the eggs in a nest built in a hollow in a tall living or dead eucalypt. The male brings her food and calls to her to be fed on a branch outside the nest.

Cockatiel
Nymphicus hollandicus

Other Names: Quarrion, Cockatoo-Parrot, Weero, Crested Parrot

Distribution: Inland parts of Australia generally, where suitable habitats exist, rarely extending to coastal areas.

Notes: The Cockatiel gets its nickname 'Cockatoo-Parrot' from the much pondered question as to which group the bird rightly belongs. Research and studies have determined the Cockatiel is a cockatoo, but it does have parrot-like wings and tail.

Cockatiels congregate in groups of several hundred to roost. They have a preference for open or lightly timbered country. They are seed-eaters and forage on the ground by gleaning the seeds of grasses, herbs and trees, especially from acacias. They drink everyday, if water is available, by scooping the liquid with the lower mandible. When Cockatiels are disturbed, they use their grey colouring as camouflage. They will fly to dead trees and perch lengthwise on the limbs. The most distinctive feature these birds have is their flight, which is graceful and buoyant on steadily beating wings. They also warble a pleasant, chattering note to each other in flight.

Blue Bonnet
Northiella haematogaster

Other Names: Bulloke Parrot, Red-bellied Blue Bonnet, Naretha Parrot, Little Blue Bonnet, Red-vented Blue Bonnet, Crimson-bellied Parrot

Distribution: Western New South Wales, north-western Victoria, and eastern South Australia; a distinctive population, the Naretha Parrot, inhabits the western fringes of the Nullarbor Plain.

Notes: There are four races of Blue Bonnet in Australia each with subtle differences, but basically the same. They prefer the semi-arid southern inland of Australia where they will be found in open casuarina, cypress pine and acacia woodlands. The name Blue Bonnet is somewhat of a misnomer because it is only the bird's face which is blue, not the head as the name implies. Blue Bonnets are closely related to Grass Parrots, yet they also have many attributes of rosellas. Their fawn-brown plumage is, however, distinctly characteristic of this bird.

Blue Bonnets are gregarious birds, forming permanent pairs which socialise and feed in flocks of up to 30. They will often be seen on the ground in the shade of a tree, feeding on the seeds of native plants. During the heat of the day they perch quietly on the foliage of trees and shrubs where their natural camouflaging can make it difficult to spot them, they feed and drink in the early morning and late afternoon.

During the breeding season, the male Blue Bonnet likes to display for its mate. It will stand erect, raise and vibrate its wings, fan and shake its tail from side to side, bob its head, and raise its crown feathers into a crest.

Turquoise Parrot
Neophema pulchella

Other Names: Turquoisine Parrot, Chestnut-shouldered Grass-parrot, Beautiful Grass Parakeet, Red-shouldered Parakeet

Distribution: From about Maryborough, Queensland south through eastern New South Wales to the Victorian border.

Notes: The Turquoise Parrot lives in pairs or small flocks, frequenting rough grasslands bordering on open forests. It spends most of the day on the ground feeding on seeds, preferably in the shade of a tree. When they are disturbed they fly swiftly with erratic fluttering and a flash of their yellow tail feathers to the nearest tree.

This pretty bird was once more common than it is now but suffered a decline around the turn of the century. Although it was once thought in danger of extinction this seems to have been overcome through the preservation of their habitat in the Warrumbungles National Park in New South Wales.

Rainbow Lorikeet
Trichoglossus haematodus

Other Names: Coconut Lory, Red-collared Lorikeet, Blue Bellied Lorikeet, Blue Mountain Lorikeet.

Distribution: Widespread in most forests and closer woods of northern and eastern Australia from the Kimberleys to Cape York Peninsula, Tasmania and the Eyre Peninsula, South Australia.

Notes: The Rainbow Lorikeet is a prime example of Lorikeets in general as it demonstrates much of the behaviour and habitats of all lorikeets. It has the auspicious label of being the first Australian bird to be illustrated in colour, in 1774, well before the British colonised New South Wales.

The lorikeet seeks out blossom laden trees where it feeds on nectar, pollen, flowers and fruits of native and cultivated fruit trees, as well as grain. Like all lorikeets its tongue is covered with brush-like papillae which help it mop up food from flowers. While feeding they clamber easily along branches and it is usual to see a lorikeet hanging upside down on a perch.

The Rainbow Lorikeet is to be found in company, often in flocks of around 50 birds. They roost at night in the tree tops, often combining with one or two other flocks chattering and preening each other. Each morning they fly off in search of food, often many kilometres, and they screech to each other in shrill, sharp calls as they fly.

Musk Lorikeet
Glossopsitta concinna

Other Names: Green Keet, Green Leek, Musk Lory, Musky Lorikeet, Red-crowned Lory, Red-eared Lory

Distribution: From Rockhampton, Queensland, to Tasmania and across to the Mount Lofty Ranges in South Australia and Kangaroo Island.

Notes: The Musk Lorikeet frequents eucalypt woodlands and drier forests in the foothills and fringes of plains. They will also be found in road side timber areas and along timbered watercourses. This bird is noted for its rolling high pitched screech and its chattrering as is it forages for food. They spend the day in the feeding trees eating nectar from flowering eucalypts, fruits, berries and seeds. Flocks will also eat ripening wheat and sorghum crops to which they can do a great deal of damage. At night they return to their roosts and pair up and preen each other.

Musk Lorikeets are nomadic birds, travelling mostly in autumn and winter. They are so swift in flight that their beating wings make a whirring noise which can be heard as the bird passes overhead. They build nests high in eucalypt trees in holes which they line with decayed debris.

Budgerigar
Melopsittacus undulatus

Other Names: Budgerygah, Betcherrygah, Shell Parrot, Shell Parakeet, Canary Parrot, Zebra Parrot, Lovebird.

Distribution: Ranging throughout most of Australia, chiefly the interior. Inhabits trees bordering watercourses, sparsely timbered grasslands, mallee, mulga and spinifex desert.

Notes: The Budgerigar is a popular "pet" being kept as a caged bird in many households. This small, pretty bird has various domesticated strains in many countries. It is a striking bird with its obvious stripes covering its face, head and upper back.

In the wild the "Budgie" lives in flocks of around 100 birds and is highly nomadic, its movement largely dependent on the availability of food. It feeds on grasses, chenopods and other herbs. When they feed on the ground they will often be seen climbing up and down tussocks of grass. Sometimes while feeding, and mostly when perched in trees, the birds warble and chatter to each other. They roost in trees along waterways and tend to change their roost site each night.

During the breeding season the male and female will chase each other a great deal before settling down on a branch where the male will wrap a wing around the female while they mate. The female builds the nest and broods unaided but all the while she is fed by the male by regurgitation.

Pacific Heron
Ardea Pacifica

Other Names: White-necked Heron

Distribution: Australia generally.

Notes: The Pacific Heron lives in pairs or small flocks inhabiting swamps, rivers or the borders of lakes. It is a shy and wary bird, rather uncommon, often appearing only during droughts. The heron feeds by wading around shallow pools of less than 10cm depth and by stalking through wet grass. They hunt from an upright or crouched position watching carefully and poised all the time ready to strike. When they spot their prey they launch out and grab it, swallowing it head first. Their diet consists of crustaceans, tadpoles, frogs, fish and aquatic insects.

When the Pacific Heron takes to the air it tucks its long neck in so its head is close to its body. It trails its long legs out behind, and flaps continually with very little gliding.

Pied Cormorant
Phalacrocorax varius

Other Names: Black and White Shag, Yellow-faced Cormorant

Distribution: Coastal mainland Australia and eastern Australia.

Notes: Pied Cormorants are the least nomadic of any Australian cormorant. They live in both fresh and salt water, but prefer open coastal and sub-coastal inlets and lakes than inland rivers. In coastal colonies they will often congregate in thousands, and prefer to breed in colonies wherever there is an adequate food supply.

During courtship the males choose a nest site and advertise by silently wing-waving for several seconds at a time. They throw back their heads, point their bills up, spread and cock their tails, and flutter their wings at two flaps a second. When a female approaches, they drop their wings and utter gutteral cries. Once pairing has taken place, the male gathers sticks and debris for the nest platform and the female builds it. They complete their bonding by passing nest material between one another, bill fencing, intertwining their necks, and mutual preening. While sitting on the nest the male will call 'wog-wog-wog' notes of greeting to his mate and the female hisses in reply.

Pied Cormorants actively fish underwater. They dive headfirst from the top of the water and propel themselves swiftly underneath with their fully webbed feet pushing in unison. They grasp their prey, such as fish, crustaceans and molluscs, in their hooked bills and then return to the surface where they eat them. To help the bird focus underwater they have a thick, transparent membrane and a special muscle to squeeze the eye's lens in an instant. Their nostrils are sealed and the birds breathe through valves at the corner of the mouth.

When they are not fishing, Pied Cormorants spend their time perched on dead trees, boats or poles, drying their wings and oiling their feathers with the help of two large preen glands.

Lotus-Bird
Irediparra gallinacea

Other Names: Comb-crested Jacana, Comb-crested Parra, Lilytrotter, Christbird, Skipper

Distribution: Coastal and subcoastal northern and eastern Australia, west to the Kimberleys and south east to the Hawkesbury River, New South Wales, where irregular.

Notes: The Christbird is a good nick-name for the Lotus-bird as its agility at walking on water weeds makes it look as though it is walking on water. The bird is able to do this because of its exceptionally long hind toe which helps distribute its weight evenly. These birds spend their whole life on the floating vegetation of swamps, lagoons and streams, foraging along the edges of pools for aquatic plants, seeds and insects.

The Lotus-bird even builds its raft-like nest on leaves or grass growing in water generally more than one metre deep. Both the male and female look after their chicks, and they become very wary when nesting. The Lotus-bird has the ability of being able to dive and swim underwater, even though they do not have webbed feet, and they can remain motionless up to half an hour under water with just their nostrils and bill above the surface. An adult bird can also carry a chick or an egg under its wing, which is vital if the water level rises or falls suddenly.

Cattle Egret
Ardeola ibis

Other Names: None

Distribution: Coastal Australia generally; common and widespread in the north and east, less so in the west and south.

Notes: Cattle Egrets were introduced from Africa to the Kimberley Region in the 1930s in the hope of controlling the cattle-tick through its habit of perching on cattle, and buffalo, and picking off insects and parasites. For this reason, and also because of the birds tendency to feed around the feet of slow-moving cattle, it is called "Cattle" Egret. The spread of the bird today suggests that extensive migration from Asia to Australia has occurred naturally, and that their numbers are not entirely due to its introduction.

Cattle Egrets forage in wet pasture land in pairs or small groups. They strut along with nodding heads and make a dash at their small prey. During the breeding season the birds form colonies of thousands, often with other herons and egrets. They make large platform nests of sticks and twigs in trees or shrubs along waterways. Males choose each nest site and defend it. They acquire orange-buff head plumes and a red bill and carry twigs in order to attract a female.

Pacific Black Duck
Anas superciliosa

Other Names: Black Duck, Wild Duck, Grey Duck, Brown Duck, Blackie

Distribution: Continental Australia generally and Tasmania, but seldom in inland deserts.

Notes: On almost every body of fresh or brackish water in Australia you are sure to find a Pacific Black Duck. Although called "black" the birds are, in fact, speckled brown with distinctive black and white stripes on the face.

A typical surface feeding Pacific Black Duck takes both plant and animal food from the water by dabbling, dredging and upending. They like, particularly, seeds of grasses, weeds and water plants, as well as crustaceans, molluscs and insects. At times flocks have been known to forage in irrigated rice fields, yet, contrary to expectations, they do very little damage to them.

Breeding occurs when water areas provide them with abundant food, this generally being in the spring in the south, and in autumn in the tropical north. Females lay their eggs in a wide variety of nests ranging from scrapes in the ground, to well-woven cups in grass or reeds, these often being the abandoned nests of other birds. When the female leaves the eggs to feed, she covers them in a layer of down to keep them warm.

Red-Necked Avocet
Recurvirostra novaehollandiae

Other Names: Australian Avocet, Cobbler, Cobbler's Awl, Trumpeter, Painted Lady, Yelper

Distribution: Australia generally, except northern and eastern coastal regions; vagrant in Tasmania.

Notes: The Red-necked Avocet can be found frequenting lakes, streams and swamps in most parts of Australia. They wade through the shallow water on their long, stilty legs with their heads down and forward, swinging their bills from side to side in search of aquatic animals and plant matter. The bird gets its name "Cobbler's Awl" from its distinctive bill which curves upwards for most of its length, but downwards at the tip. It is also this bill which helps distinguish the bird from its close relative the stilt.

The Avocet is usually a shy, quiet bird but becomes noisy if its nest is approached, at which time it will call out like a yelping puppy. At other times, usually when in flight, the bird sounds a soft, trumpeting whistle.

They are a nomadic bird, following heavy rainfalls, and breed from August to December. While courting the birds will display by bowing, trampling, dipping their bills and preening. The nest tends to be no more than a depression in the ground lined with scraps of swamp vegetation. Here the female lays her four eggs.

Rufous Night-Heron
Nycticorax caledonicus

Other Names: Nankeen Night Heron, Night Heron

Distribution: Australia generally.

Notes: The Rufous Night-heron is mainly a nocturnal bird, frequenting wetlands generally. By day they camp in groups under the cover of leafy trees, especially weeping willows, or in leaf beds. Towards dusk they emerge from cover croaking sporadically, and feed along the water's edge for freshwater molluscs, fish, frogs and aquatic insects. Rufous Night-herons have also been known to eat the chicks and eggs from other birds' nests, and a high percentage of them carry the virus for Murray Valley encephalitis.

During the breeding season, males and females come together into large colonies. The males stake out small nesting territories and begin a dance display in which they tread with their feet and jerk their heads down to their toes, raising crest and nape plumes with accompanying noises. The chicks are fed by both parents by regurgitation, and later they cough food into the nest for the chicks to pick up.

Wood Duck
Chenonetta jubata

Other Names: Australian Wood Duck, Blue Duck, Maned Duck, Maned Goose

Distribution: Throughout Australia generally.

Notes: The Wood Duck can be found in lightly timbered and grassy areas as well as on the margins of swamps, rivers and lakes, mainly in the tablelands and the western slopes of eastern Australia. Although they are built for the water they are most often found on solid ground and not swimming. They are nomadic birds and follow the rain from place to place.

These shy, wary ducks are usually found in very large flocks which disperse in the summer months only to come together again in the winter. Generally they will return to the same area year after year, and to the same mate. When these birds are disturbed they will freeze, or flatten themselves along the ground to resemble dead wood. They can run quite swiftly and agilely but are clumsy and awkward in the water, which is unusual for a duck. They eat leaves, seeds of grasses, cloves and sedges, on the land and in the water, often feeding many kilometres from where they roost at night.

Masked Lapwing
Vanellus miles

Other Names: Masked Plover, Spur-winged Plover, Wattled Plover, Alarmbird

Distribution: Australia generally, except the centre and south-west.

Notes: The Masked and the Spur-winged Plover were previously thought to be two distinct species, but now they are regarded as different races of the one species — Masked Lapwings. They frequent wet lands in general including lush pasture land, marshy ground, golf courses and city parks, rarely being very far from water. Here they stalk watchfully, shoulders hunched and head forward. When they spot their prey, usually a spider, small crustacean, worm or insect, they dart their bill down and grasp it.

The Masked Lapwing is a fairly sedentary bird, rarely travelling far between its roosting and feeding grounds. At night and during the day they roost or rest standing in shallow water on small islands. They call to each other in their strident, staccato "keerk, keerk, keerk" when alarmed or in flight.

Black-Winged Stilt
Himantopus himantopus

Other Names: Pied stilt, Whiteheaded Stilt, Long-legged Plover, Longshanks, Dogbird, Stiltbird

Distribution: Australia generally, including Tasmania; local and patchy over much of the arid interior.

Notes: One of the most striking features of the Black-winged Stilt is its long, pink stilty legs, which, at about 20cm in length, make them the longest legs on a bird in proportion to its body size. While these long legs are ideal for wading in wet mud and shallow waters they prevent the bird from swimming or submerging in search of food. This shy, black and white bird may be seen in small flocks in any swamp area, lake or river throughout Australia. There it will be stalking for food, jabbing the mud with its needle-like bill. It feeds on molluscs, flies, aquatic insects shrimps or plant matter.

Black-winged Stilts are nomadic birds, moving on to new areas as water levels change. They fly with their legs trailed far behind them using them like a rudder to manoeuvre, while their short-pointed triangular wings beat rapidly and shallowly. They are also a wary bird and take quickly to the air when disturbed, sounding a yelping bark as they move off.

Sacred Ibis
Threskiornis aethiopica

Other Names: White Ibis, Black-necked Ibis, Sickle-bird

Distribution: Eastern Australia generally, from the Kimberleys south to the Eyre Peninsula.

Notes: The Sacred Ibis is marked by having no feathers on its black head and neck; here it is completely bald. Like all Ibises, it also has the distinctive curved and very sensitive bill which it uses when feeding by probing in swampy or water covered ground. It continually moves its head from side to side in the search for freshwater crayfish, water insects, fish, snails and small snakes. Occasionally the ibis will hold its head and upper neck under water when feeding.

A flock of Sacred Ibis flying is a wonderful sight. They fly in a stepped or V-formation, each bird flapping its wings at the same time as all the others. At times they will soar upwards on warm, thermal air currents up to 3000 metres. Back on the ground a pair of mating Sacred Ibises will begin their courtship displays. The male secures a display territory and prepares a nest site. He then signals his welcome to the female from his perch through a series of bows. The female lands close to him showing her interest and from there the male leads her to the nest site. He grasps a twig in his bill and repeats his bowing until the female grasps the twig also.

Straw-Necked Ibis
Threskiornis spinicollis

Other Names: Dryweather-bird, Farmer's Friend, Letter-bird

Distribution: The Australian mainland generally; migratory only to Tasmania.

Notes: The Straw-necked Ibis is a slender bird with a long down-curved bill and stiff straw-like feathers on its throat, giving it its name. They frequent swamps, the margins of streams and lakes, and pastoral lands. They feed in both wet and dry pastures eating water insects, molluscs, frogs, snakes, caterpillars and grasshoppers. Often they will follow a plague of locusts or other insects and will devour large numbers of them.

During the breeding season, large colonies of hundreds of thousands of Straw-necked Ibises will gather in search of nesting islands and swamp thickets. During courtship a red patch of skin develops behind the eyes and on each side of the breast. This soon fades after pairing is completed. Courtship displays of bowing and mutual preening take place at the nesting site. The pair build the nest together with the male bringing the sticks and the female pushing them into the nest. When the eggs are laid the pair will incubate in shifts, bowing to each other when changing over. The pair also bow to the young before they regurgitate food for them to eat.

Yellow-Billed Spoonbill
Platalea flavipes

Other Names: Yellow-legged Spoonbill

Distribution: Throughout most of Australia though apparently absent from Cape York Peninsula, central Australia, and the inland parts of Western Australia.

Notes: Yellow-billed Spoonbills inhabit freshwater swamps, margins of lakes and streams, and mud covered flats. It can usually be found stalking the edges of a swamp in search of food. They feed in the shallows by slowly moving their head in a sweeping movement. They let the water run through their partly opened mouth and if anything touches the inside of the spoon-shaped tip they use the knobs in the bill to filter it further into the mouth. They also use their bills and feet to stir the debris and organisms from the water bottom. Yellow-billed Spoonbills eat water insects, crustaceans, fish and molluscs.

Yellow-billed Spoonbills are fairly common, but are highly nomadic and erratic in occurrence. They return to the same breeding ground each year as long as there is water. In their courtship behaviour the female spoonbill can be very aggressive towards the male until she accepts him. This bond is made when the female allows the male to nibble her bill.

Pied Heron
Ardea picata

Other Names: Pied Egret, White-headed Egret

Distribution: Tropical northern Australia from the Fitzroy River to the Gulf of Carpentaria and east to the Burdekin System.

Notes: Pied Herons are gregarious birds, living in loose flocks of 5-30 birds, in shallow freshwater or salt swamps, wet grasslands, lagoons and mangroves in the tropical north. They feed on crustaceans, small molluscs, fish and insects and their larvae. They capture their prey by walking quickly through low and floating vegetation, gleaning it with rapid and repeated pecks. Unlike other herons they tend not to stand and watch, preferring to peck frequently. Sometimes, in deeper water, they will stand and wait for their prey lunging at it at the optimum moment from either an upright or crouched position.

At times Pied Herons will be found in flocks of a hundred birds or more. This is likely to be in the breeding season because they nest in colonies, and often with other species like the Intermediate Egret and Little Pied Cormorant. In their nesting territory the heron will build a slightly concave nest of twigs and sticks in a mangrove tree, about 5 metres above the ground or water.

Buff-Banded Rail
Gallirallus philippensis

Other Names: Banded Land Rail, Land Rail, Painted Rail, Little Tarler Bird, Pectoral Rail, Corncrake

Distribution: Coastal north, east, south-east and west Australia; also isolated inland areas.

Notes: The Buff-banded Rail lives in coastal and subcoastal areas, rarely venturing far inland. Its habitat is dense, tussocky vegetation and shrubberies around swamps and lagoons, among mangroves, along watercourses and on islands. It is a rarely seen bird because its colouring provides excellent camouflage amongst the vegetation.

The Buff-banded Rail feeds on aquatic vegetation, seeds, insects and small molluscs. Occasionally it will venture into areas where domestic fowl feed. It is a wary bird, however, and will dash into dense cover at the least disturbance, flicking its tail in the characteristic manner of rails. Also, when it is startled, it utters an alarm "krek" which is often answered by other birds in the vicinity. In their breeding season, from September to January, the Buff-banded Rail builds a nest of grass or reeds pulled down and woven into a saucer shape and placed on a tussock. Here the female will lay as many as eight eggs.

Dusky Moorhen
Gallinula tenebrosa

Other Names: Black Gallinule, Black Moorhen

Distribution: Eastern Australia north to the foot of Cape York Peninsula, far southwestern Australia and northeastern Tasmania.

Notes: The Dusky Moorhen lives in freshwater swamps and parkland lakes in eastern Australia where it feeds by day in fresh water by swimming and diving for food, or on land by walking. It prefers land and aquatic plant matter as well as fish, molluscs, worms and insects. They are sedentary birds, and very territorial, fighting aggressively to defend their areas.

They form family groups of between two and seven birds made up of one to three males to each female. Together they drink and bathe, preening each other. They spend nights together on roosting platforms up to 2m above the water in reeds and shrubs. They sleep standing up. The Dusky Moorhen has the unusual habit of having its legs turn red during mating. Females initiate courtship and will mate with all the males in their group. Together, all of them will build a slightly dished platform nest of aquatic vegetation, usually among rushes or at the base of a tree growing in a swamp. The female lays her eggs daily until the clutch of five to eight eggs is complete, then the group take turns sitting them. When they have hatched, the young are nursed for about three days in a second nest over deeper water until they are ready to go out foraging with an adult bird. The chicks remain under parental care for around nine weeks.

Darter
Anhinga melanogaster

Other Names: Snake-bird, Diver, Eastern Darter, Oriental Darter, Needle-beak Shag

Distribution: Mainland Australia generally.

Notes: The Darter lives on inland lakes, streams and swamps with its favourite haunts being deep pools and secluded reaches. Here it hunts for prey — small fish, insects and aquatic animals — by submerging its body under water, leaving only its head and neck showing. It makes snake-like movements to and fro as it watches for fish, hence its nickname, Snake-bird. It may pursue the fish underwater, or it may stalk its prey and wait for it to come close. When it is within range the bird holds back its long S-shaped neck, then suddenly darts out and spears its victim with its dagger-like bill.

Unlike most birds, Darters do not have water proof feathers. Instead they use their waterlogged feathers to lower their buoyancy when they are hunting. To dry their feathers the birds remove the water by squeezing it out of their plumage with their bills, and they will often sit with their wings outspread to dry them. They also use oil from their enlarged preen gland, situated at the base of their tail, to help repel water.

Darters leap from their perches to take off into flight. They make a few flapping motions and then glide, but will often soar up in spirals, on thermal air currents, with ibises and pelicans. They have a large, fan-shaped tail which silhouettes characteristically.

Pacific Gull
Larus pacificus

Other Names: Jack Gull, Molly Gull

Distribution: Breeds on islands off the southern coast of Australia; occurs north to about Shark Bay in the west and about Sydney in the East.

Notes: The Pacific Gull can usually be found singly or in pairs, and occasionally loose groups. It is very much a shoreline bird, patrolling tidelines and rarely straying inland farther than saline coastal lakes. The adults are sedentary, selecting and maintaining the same stretch of beach all year. The immature birds, however, tend to wander, often travelling hundreds of kilometres before returning to establish themselves in a breeding area.

The Pacific Gull has a larger and heavier bill than any other gull, and they use this to their advantage when feeding. They fed on fish, squid, crabs, molluscs, sea urchins, and are notorious for robbing the nests of other seabirds. They "crack open" shelled prey by dropping them from a height onto rocks below. Occasionally they will forage rubbish dumps for food.

The Pacific Gull has a graceful and leisurely flight pattern on deep and slow beating wings. They rest by perching on spits, reefs, beaches and wharves.

Sooty Tern
Sterna fuscata

Other Names: Egg-bird, Whale-bird, Wideawake

Distribution: Occurs only in the tropical areas around the east and west coasts of Australia. Breeds only on islands.

Notes: The Sooty Tern is a rare visitor to Australian shores usually only being found after the storms and cyclones on the west and east coasts, in the northern part of the continent. They are a very distinctively marked bird which helps distinguish them from other terns. It is the only wholly sooty grey tern to have a forked tail, and it also has prominent pale crescents on its mantle and spots on its wing. The bird has a high pitched call of 'ker-wack-wack' which led sailors to give the bird the nickname of 'wideawake', which was what you were when you heard these birds cry! They are almost exclusively an aerial bird settling rarely on land or water, except for breeding.

At first, Sooty Terns were known only to sailors and fishermen who visited the remote islands in the tropical seas on which the birds breed. Today they still form large colonies on these islands and scrape nests in the sand, out in the open or under a bush, and lay their single egg. They will not breed until they are at least four years old, but more likely somewhere between six and eight years. The birds seem to feed by skimming the surface of the sea for plankton, small fish, squid and crustaceans, but they will also dive into the water to get their catch of the day.

Silver Gull

Larus novaehollandiae

Other Names: Seagull, Red-billed Gull, Red-legged Gull, Mackerel Gull, Kitty Gull, Jameson Gull.

Distribution: All Australian coasts, inland waters and urban parks.

Notes: The Silver Gull, or Seagull as it is best known, is a common sight on Australian beaches as it scavenges for food, even to the point of pestering humans. They are also often seen in flocks in city parks, picnic areas, rubbish dumps, on rivers and bays. They feed on fish, plankton, crustaceans and aquatic insects. After heavy rain they frequent flooded ovals to feed on worms that are forced to the surface, and enjoy feasting on insects in newly-plowed fields.

Any group of gulls establishes a hierarchy of dominance with older birds prevailing over younger ones. This is often seen in the aggressive displays when fighting for food. They hunch, arch and run forward squawking, in a variety of aggressive postures, or beg in submission.

Silver Gulls breed in colonies on islands and return to the same area each year and usually to the same mate. They build shallow nests of fine plant matter, generally on the ground and occasionally in trees or bushes. While the chicks are growing primary wing feathers, they shelter under plants within the nest territory. If they wander from their territory too early they run the risk of being killed by other gulls.

White-capped Noddy

Anous minutus

Other Names: Black Noddy, Titerack

Distribution: Nests on islands and coral cays from Torres Strait south to the Capricorn Group, Queensland.

Notes: The White-capped Noddy is very similar to the Lesser Noddy but its forked tail makes it quite distinguishable. It is also very like the Common Noddy but has a longer, more slender bill. These noddies are not migratory birds but rather roost or nest in large colonies on coral islands where their deafening calls can sometimes be heard up to two kilometres out to sea.

During the day the birds leave their nesting and roosting grounds and head out to the open sea to feed. They pick fish, cuttlefish, other molluscs, jelly fish and small plankton from the surface of the water but will also dive for them. In the evening they will return to feed their young, who also disappeared during the day but have since returned. White-capped Noddies are so tame they can be easily lifted from the nest which may be built up to five metres from the ground. The nests are interesting constructions of twigs, leaves and seaweed all bound together with droppings. When the birds are not in their nests they may be seen sunbathing with their tails and one wing spread, and their heads to one side.

Little Penguin
Eudyptula minor

Other Names: Fairy Penguin, Blue Penguin

Distribution: Coasts of southern Australia from about Fremantle, Western Australia to about Port Stephens, New South Wales.

Notes: The Little, or Fairy Penguins are one of Victoria's most famous tourist attractions, as the parade every night of birds returning from the sea to their nests attracts many sightseers. They are the smallest of all penguins, the only ones which breed in Australia, and the only ones to wait until dark before climbing out of the sea to waddle off to their burrows.

These penguins are very much at home in the sea. They are strong, fast swimmers using their modified wings, or flippers, to propel them through the water and help them dive and turn while they pursue fish. Their bodies are torpedo-shaped, with thick necks, and they are covered in short, stiff body feathers. Underneath, thick down covers the body and traps air to provide insulation.

Early in the morning the penguins leave their burrows in a reversal of the previous evening's arrival. Just before dawn they yap to each other in their sharp barking tone, much like a small dog. Then they emerge from their burrows and head down to the beach where they plunge into the water. Here they spend the day feeding on fish and squid.

Index

Acknowledgements

I am indebted to the assistance given from Kevin Mason, Ian McCann, Alan Gibb and Jim Pickford, also to the management and staff of the Healesville Sanctuary and the Melbourne Zoo in the production of this book.

The equipment used was Hasselblad E.L.M., Nikon 801 with Kodachrome and Fuji films.